GREEN CHEMISTRY

Edited by **Hosam El-Din M. Saleh**
and **Martin Koller**

Green Chemistry
http://dx.doi.org/10.5772/68007
Edited by Hosam El-Din M. Saleh and Martin Koller

Contributors

Hosam El-Din M. Saleh, Martin Koller, Shagufta Perveen, Sandra Duque Gaona, Siti Zullaikah, Orchidea Rachmaniah, Adi Utomo, Yi Hsu Ju, Helda Niawanti, Mumtaz Quraishi, Chandrabhan Verma, Eno Ebenso, Nurettin Menges, Keyla Soto, Carlos R Cabrera, Muhammad Shahzad Aslam, Sarwat Raja, Muhammad Syarhabil Ahmad, Sachin Pandurang Shirame, Raghunath Bhosale

Published by InTech
Janeza Trdine 9, 51000 Rijeka, Croatia

Publishing Process Manager Nina Kalinic
Technical Editor SPi Global
Cover InTech Design team

First published February, 2018
Printed in Croatia
Legal deposit, Croatia: National and University Library in Zagreb

Additional hard copies can be obtained from orders@intechopen.com

Green Chemistry, Edited by Hosam El-Din M. Saleh and Martin Koller
p. cm.
Print ISBN 978-953-51-3847-1
Online ISBN 978-953-51-3848-8

For EU product safety concerns:
IN TECH d.o.o., Prolaz Marije Krucifikse Kozulić 3, 51000 Rijeka, Croatia,
info@intechopen.com or visit our website at intechopen.com.

Meet the editors

Hosam El-Din M. Saleh is a professor of radioactive waste management at the Department of Radioisotope, Atomic Energy Authority, Egypt. He received his MSc and PhD degrees in Physical Chemistry from the Cairo University. He has also been interested in studying innovative economic and environment-friendly techniques for management of hazardous and radioactive wastes. He has authored many peer-reviewed scientific papers and chapters. He is the editor of different books related to valuable international publishers.

Martin Koller is an experienced senior researcher in the field of biomediated polyhydroxyalkaoate (PHA) production, encompassing the design and development of continuous and discontinuous fermentation processes and novel downstream processing techniques for sustainable biopolymer recovery from microbial biomass. His research focus is on enhanced cost-efficiency of PHA production from surplus materials using both eubacteria and halophile archaea as whole cell biocatalysts.

Contents

Preface

Currently, the expression "Green Chemistry" is an omnipresent subject matter in both academic and innovative chemical companies. All over the world, efforts are dedicated to initiatives to make chemical processes of manufacturing more sustainable, more energy efficient, more biocompatible, more environmentally benign, less waste-generating, healthier, and more resource-efficient; all these issues nowadays are jointly discussed as "Green Chemistry."

Today, the modern chemist is aware of the fact that research and development always have to meet the ancient Latin principle "Quidquid agis, prudenter agas, et respice fines!" ("whatever you do, do it virtuously, and consider the consequences"). This principle implies that every new process in the development has to take into account the possible impacts on the current and future status of humankind, the environment, and, most of all, the fate of future generations. Translated into our present world, chemists and engineers should act in accordance with the 12 well-known principles of Anastas and Warner, which honestly can be regarded as the basic law of current and future chemical manufacturing practice. In a nutshell, these 12 principles show us the way chemistry in general can get rid of its bad public reputation, which brands it as "toxic," "polluting," "corrosive," "lethal," "stinking," and so on. In this context, Anastas and Warner postulated that chemistry can definitely become "green," when fundamental paradigm shifts are implemented in its manufacturing processes. This includes switching from exploiting limited resources of fossil origin to profiting from the abundance of renewable feedstocks and the development of atom-economic processes, generating "zero waste," producing materials in an intrinsically safe way, designing energy-efficient processes, or using harmless catalysts. Especially the last aspect, namely the use of benign and efficient catalysts, provides the direct link between "green chemistry" and the nature's approaches of chemical manufacturing; here, "biocatalysts," hence enzymes, are used to efficiently, selectively, and sustainably transform diverse raw materials into desired marketable end-products. After completing their task, enzymes undergo biodegradation within the nature's closed material cycles. Regarding strategies for producing materials in an intrinsically safe way, one can refer to the emerging field of "flow chemistry" using advanced microreactor systems, which allow carrying out typically highly hazardous reactions efficiently and safely.

The book at hand, "Green Chemistry," was compiled to mirror all the upcoming discussed aspects, which will make chemistry more sustainable. Carefully selected authorships from different countries contributed well-chosen, specific chapters to this publication endeavor. The introduction section written by Hosam Saleh and Martin Koller provides the embedding of different chemical techniques into the principles defined by Anastas and Warner in order to provide a fundamental overview on what "Green Chemistry" should aim at. Quraishi Mumtaz presents the new aspects of using ionic liquids as "green" corrosion inhibitors to protect industrial metals and alloys. Perveen Shagufta makes us familiar with the question how "green chemistry" can be implemented in the production of desired anticancer molecules. Moreover, "green" approaches for new separation techniques to recover bioactive natural products are described by Zullaikah Siti. Soto Keyla focuses on the utilization of zero-valent iron nanoparticles for the removal of cadmium and different other metal ions from polluted environments. "Green" approaches in "click chemistry" to synthesize derivatives of 1,2,3-triazol-1-yl piperazine, 1,2,3-triazol-1-yl quinoxaline, one pot 1,2,3-triazole and bistriazole, important bioactives in antimicrobial, antioxidant, anticancer, antiviral, anti-HIV, and antitubercular research are summarized by Shirame Sachin. Gaona Sandra reports on the implementation of "green chemistry" metrics on the educational level, whereas Menges Nurettin focuses on the use of "green" catalysts and environmentally benign solvents in chemical manufacturing. A completely solvent-free process to extract essential oils is presented by Aslam Muhammad Shahzad. Finally, Sandra Duque Gaona makes us familiar with the utility of the toxic release inventory (TRI) in tracking implementation and environmental impact of industrial "green chemistry" practices in the United States.

With this issue, the editors wish to address scientists active in the diverse fields of "green chemistry," encompassing chemistry, chemical engineering, and environmental science. The book is also dedicated to students of higher level who are involved in these fields; we hope that this book is helpful and provides inspiration for readers during their studies and later in their academic or industrial career! In addition, we strongly believe that the issue also attracts the attention of representatives of the chemical industry. Do you, as an industrial responsible, want to get to feet on the ground of sustainable manufacturing processes? This might provide for the ignition sparks for a broader implementation of "green chemistry" on an industrial scale.

We are tremendously optimistic that the exploratory and scientific efforts collected and summarized in the book at hand will motivate researchers all over the planet to deepen their R&D activities in this field and to attract the interest of undergraduates as well as of innovative representatives from relevant industrial sectors. Primarily, these activities shall boost the impatiently desired breakthrough of "green" manufacturing processes, which fairly merit this designation. As certified by the rich content of this book, sustainable, often even "bio-inspired," remedies for prevalent ecological problems are already available, developed by experts from different scientific fields, or these solutions are at least in a well-advanced stage of development; they are expecting their industrial implementation in the emerging field of "green chemistry!"

In particular, I would like to acknowledge the Publishing Process Manager, Ms. Nina Kalinić, for her prosperous cooperation, exceptional efforts, and prompt response to my requests. Again, we would like to thank cordially all the contributors to this issue for their supreme work.

Editor: Hosam El-Din M. Saleh, Ph.D.
Atomic Energy Authority
Radioisotope Department, Nuclear Research Center
Giza, Egypt

Co-Editor: Dipl.-Ing. Dr. Martin Koller
University of Graz
Office of Research Management and Service, c/o Institute of Chemistry
Graz, Austria

Section 1

Principals of Green Chemistry

Chapter 1

Introductory Chapter: Principles of Green Chemistry

Hosam El-Din Mostafa Saleh and M. Koller

Additional information is available at the end of the chapter

http://dx.doi.org/10.5772/intechopen.71191

1. Introduction

New chemistry is required to improve the economics of chemical manufacturing and to enhance the environmental protection. The green chemistry concept presents an attractive technology to chemists, researchers, and industrialists for innovative chemistry research and applications.

Primarily, green chemistry is characterized as reduction of the environmental damage accompanied by the production of materials and respective minimization and proper disposal of wastes generated during different chemical processes. According to another definition, green chemistry is a new technique devoted to the synthesis, processing, and application of chemical materials in such manner as to minimize hazards to humankind and the environment.

Numerous new terms have been introduced associated with the concept of "green chemistry," such as "eco-efficiency," "sustainable chemistry," "atom efficiency" or "atom economy," "process intensification and integration," "inherent safety," "product life cycle analysis," "ionic liquids," "alternate feedstocks," and "renewable energy sources."

Hence, there is an essential need to improve the synthetic and engineering chemistry either by environmental friendly starting materials or by properly designing novel synthesis routes that reduce the use and generation of toxic substances by using modern energy sources.

2. Basic principles of green chemistry

Green chemistry is generally based on the 12 principles proposed by Anastas and Warner [1]. Nowadays, these 12 principles of green chemistry are considered the fundaments to contribute to sustainable development. The principles comprise instructions to implement new chemical products, new synthesis, and new processes as illustrated in **Table 1**.

1	**The "better to prevent than to cure" principle** It is beneficial to a priori prevent the generation of waste instead of later on treating and cleaning up waste
2	**The "atom economy" principle** Synthetic production routes have to be planned in a way maximizing the incorporation of all the compounds used in the synthesis into the desired product
3	**The "less precarious chemical syntheses" principle** Wherever feasible, such synthetic methods have to be aspired, which resort to and generate compounds of no or only insignificant noxiousness to the environment and human health
4	**The "designing safer chemicals" principle** Chemicals should be developed in a way affecting their desired functionality, while, at the same time, considerably reducing their toxicity
5	**The "safer solvents and safer auxiliaries" principle** Expenditure of auxiliary substances, such as solvents, separation agents, and others, should be avoided wherever possible; if not possible, harmless auxiliaries should be used
6	**The "design for energy efficiency" principle** The environmental and economic impact of energy demands for chemical processes should be analyzed in terms of followed by optimizing the required energy input. Wherever practicable, chemical synthesis should be carried out under mild process conditions, hence, at ambient temperature and pressure
7	**The "renewable feedstocks" principle** Whenever feasible in technological and economic terms, synthetic processes should resort to such raw materials and feedstocks, which are renewable rather than limited
8	**The "derivative reduction" principle** Redundant derivatization, e.g., protection/deprotection, the use of blocking groups, or temporary modification of physical/chemical processes, requires additional reagents and often contributes to additional waste generation. Therefore, wherever possible, they should be avoided or reduced to a minimum
9	**The "catalysis" principle** Generally, catalytic reagents are intrinsically superior to stoichiometric reagents; these catalysts should be as selective as possible
10	**The "degradation" principle** Chemical products have to be designed in such a way that, at the end of their life span, they do not resist in the biosphere, but disintegrate into nontoxic degradation products
11	**The "real-time analysis for pollution prevention" principle** Advanced analytical methods have to be developed, which permit the real-time, in-line process monitoring and control well before hazardous substances are generated
12	**The "accident prevention by inherently safer chemistry" principle** Compounds and the compound's formula applied in a chemical process should be chosen in a way minimizing the risk of chemical accidents, encompassing the release of chemicals, detonations, or fire formation

Table 1. The 12 principles of green chemistry proposed by Anastas and Warner (based on [1]).

2.1. It is better to prevent waste than to treat or clean up waste after its formation

This statement is one of the most popular guidelines in process optimization; it describes the ability of chemists to redesign chemical transformations in order to minimize the generation of hazardous waste as an important step toward pollution prevention. By preventing waste generation, the hazards associated with waste storage, transportation, and treatment could be minimized.

This principle is easy to understand and easy to apply, and examples from both industry and academia have proven its significance, relevance, and feasibility. This pillar of green chemistry is still valid; however, we have to conceive it in a broader context, switching from a restricted interpretation of waste based on its quantity to a universal approach to deal with the topic "waste": (1) We have to take waste's multidimensional nature into account. (2) We need to move from a "quantity of waste per quantity of product" principle toward a principle addressing the "quantity of waste generated per function provided by the product." In this context, we have to aim at making both quality and functionality of products superior. (3) Considering the entire life cycle of a product, we have to address the fact that not only the production process itself generates waste but, moreover, "end-of-life waste" accrues after the product's life span or its consumption. This encompasses firstly the conversion of such materials up to now considered as waste into valuable products and, secondly, their recyclability.

Generally, moving toward "zero-waste production" and "waste prevention" encompasses the modernization of industrial processes through clean production techniques. These techniques aim at the reduction of gaseous emissions, effluents, solid residues, and noise generation; generally, they are developed to contribute to the protection of climate and environment [2]. However, the most auspicious strategy to prevent waste generation would simply be not producing the desired product. In most scenarios, this will not be practicable; however, it might be reasonable to instead produce completely novel products, which display higher quality and longer durability. Lower quantities of such novel, superior products are sufficient to fulfill a desired function. An alternative approach is to avoid that the product can be transformed into precarious waste, e.g., by making plastics accessible toward biodegradation or by a priori switching toward biodegradable plastic instead of highly recalcitrant petrochemical plastics. According to these ideas, we need to fundamentally reconsider our understanding of waste as hazardous material that needs to be disposed by enhancing the status of waste to a valued resource, which can act as starting material for generation of new products [3].

2.2. Maximize atom economy

Atom economy is a concept developed in the early 1990s to evaluate the efficacy of chemical conversions on an element-by-element basis [4]. In analogy to well-established yield calculations, the concept of "atom economy" is based on the ratio of the entire mass of atoms in the target product to the entire mass of atoms in the starting materials. One option to reduce waste generation is to plan such chemical transformations, which maximize the integration of all materials used in the process into the final product, resulting in a number of wasted atoms as low as possible. Hence, selecting such chemical conversion routes, which incorporate the major share of starting materials into desired products, displays higher efficiency and contributes to waste reduction. This concept is nowadays widely implemented in new routes to generate various organic compounds, e.g., such substances that are used in the biomedical and pharmaceutical field [5–7].

One factor that is greatly speeding the incorporation of pollution prevention into industrial manufacturing processes is the development of green chemistry. According to an alternative definition, chemical synthesis methods should be designed in such a way to maximize the incorporation of different potentially hazardous materials, such as spinning precarious waste with cement and sand to produce improved paste used in construction applications [8–9] or to incorporate radioactive wastes as immobilizing material to produce a safe stabilized form of waste [10, 11]. In a similar vein, combining recycled poly(ethylene terephthalate) (PET) waste with cement paste displays a viable strategy for immobilization of hazardous wastes, e.g., radioactive borates [12, 13].

2.3. Design less hazardous chemical synthesis

In synthetic organic chemistry, effecting a successful chemical transformation in a new way or with a new molecule or in a new order is what matters regarding the principles of green chemistry. Various researchers have clearly demonstrated the direct relation of toxicity and the associated hazards and risks allied with chemical reactions to the matrix of matter present in the reaction vessel. Generally, the holistic toxicity spectrum of products or processes, together with most other sustainability and green chemistry criteria, is highly impacted by the chemistry behind a process and the transformation contributing to a chemical synthesis chain. An exception is identified in such cases where a molecule is produced by purpose, which is designed to display toxicity and/or biologically activity. For example, this scenario is found in the case of various molecules synthesized for pharmaceutical or agricultural applications; such compounds exhibit toxicity and/or impact living organisms.

Selection of compounds and materials to be used to increase the efficacy of chemical transformations is a pivotal point in process development; chemists should dedicate increased attention to the decision on which materials to be put into reaction vessels. It is simple to disregard all the other materials and to dedicate all efforts exclusively to the chemosynthetic pathway, which provides us with the desired product. However, discounting all the other matter present in a production process ultimately results in a high price to be payed, and we finally have to get rid of this scenario. Sometimes, chemists actually produce hazardous molecules, and, therefore, the subsequent principle is dedicated to the design of molecules which are intrinsically safer in their nature [14].

2.4. Design safer chemicals and products

Chemical products should be designed to achieve their desired function with at the same time minimizing their toxicity. New products can be designed that are inherently safer, while highly effective for the target application. For example, the direct incorporation of radioactive spent liquid scintillation waste into cement combined with clay materials is considered an added value in the immobilization of the hazardous organic wastes in very cheap materials and natural clay to produce a safe stabilized product easy for handling, transformation, and disposal [15, 16].

2.5. Safer solvents and auxiliaries

This principle promotes the use of safer solvents and auxiliaries. It is about any substances that do not directly contribute to the structure of the reaction product but are still necessary for the chemical reaction or process to occur. Mostly, reactions of organic compounds take place in liquid milieus, where the solvent acts in different ways: it can enable enhanced contact between the reactants, it can stabilize or destabilize generated intermediates, or it can influence transition states. In addition, the applied solvent also governs the selection of adequate downstream and regeneration processes and recycling or discarding techniques. By taking the ecological effect of chemical processes in consideration, innovative concepts for substitution of volatile organic solvents have become a great challenge in green chemistry. A green solvent should meet numerous criteria such as low toxicity, nonflammability, nonmutagenicity, nonvolatility, and widespread availability among others. Moreover, these green solvents have to be cheap and easy to handle and recycle [17, 18]. Prime examples are provided in the field of extractive recovery of microbial polyhydroxyalkanoate (PHA) biopolymers, typical intracellular storage materials, from biomass [19]. For this extraction process, which is typically carried out by the use of precarious halogenated solvents, one more and more resorts to less harmful greener solvents [20–22], or to new recovery methods which entirely do without any solvents [23].

2.6. Design for energy efficiency

Usually, energy is used to enhance the human life in important ways. The traditionally used energy sources like coal, oil, and gas are limited in supply, and their combustion releases greenhouse gases. For continuous improvement of life quality, both move toward renewable energy and design for energy efficiency are needed. Designing more efficient processes by choosing the most suitable technologies and unit operations has to go in parallel with selecting proper energy sources. Using an electric motor with energy sources generated from the sun and wind is more effective in ecological terms instead of using fossil fuels. How energy is converted to useful forms and where it gets lost are the most important questions for engineers and designers to help society use energy more effectively.

Consequently, green chemistry includes minimization of energy loss like mechanical friction, fluid drag, and unwanted heat transfer, by improving the layout and insulation of a refrigerator or designing lighter vehicles with enhanced aerodynamic characteristics and lower rolling resistance.

In addition, when developing a new production process, the effect of geographical location of production plants has to be taken into account: ecological comparison of different production scenarios for the same product, in this case bioplastics, clearly demonstrates that different energy production technologies, resources for energy production, and the effect of available energy mixes in different countries become significant for the ecological footprint of a new process [24].

2.7. Use of renewable feedstocks

According to the principles of green chemistry, a raw material or feedstock should be renewable rather than depleting whenever technically and economically practicable. Using renewable resources like microbial or plant biomass, which are embedded into nature's closed carbon cycle, represents a real option to prepare functional bioproducts in a sustainable way and to contribute to energetic transition.

In the context of the Green Chemistry Principle #7, which addresses the renewable feedstock thematic, we nowadays witness a vast number of current multidisciplinary collaborations, involving the fields of, among others, agronomy, biochemical engineering, biotechnology, chemistry, microbiology, physics, toxicology, or engineering. These collaborations result in the development of next-generation fuels, polymers, and other materials pivotal for our today's society based on renewable resources and characterized by low impact on health and environment. The current global dynamic of these developments indeed gives reason to optimism for the future [25]. Finding a method to convert raw wastes such as spinney waste fibers into a mortar composite stabilized material could be an excellent application of this principal of green chemistry [8–11]. Whenever switching from fossil feedstocks to renewables, one has to consider that using renewable resources enlarges the process concept by incorporating resource provision, transportation, storage, and other aspects of logistics into the process design. Such a switch in feedstock provision, however, results in a fundamental change in the structure of processes, used technologies, and the economical concepts of industry and society [26].

2.8. Reduce derivatives

Many processes could be designed in such a way to reduce the use of additional reagents and the resulting wastes. It is commonly necessary to synthesize a derivative of a compound containing groups which are not needed in the final product, but which allow the synthesis or purification steps to proceed more easily. However, these derivatives result in lower atom economy, since they introduce atoms that are not incorporated into the final product, but finally end up as waste; this is in conflict with the atom efficiency principle according to **Table** 1. For many reactions that have traditionally required protecting groups, chemists are currently devoting research efforts to finding alternatives that do without them [27].

2.9. Catalysis

Catalysis is the chemical reaction enabled or accelerated by a catalyst. According to Ostwald, catalysts are substances that speed up a reaction by enabling an energetically favored transition state between reactants, but which are not consumed by it and do not appear in the net reaction equation [28]. Catalysts play an essential role in our modern industrial economy, in our stewardship of the environment, and in all biological processes. Saleh and others found that iron and copper sulfate as catalysts improved the mechanism of oxidative degradation of cellulosic wastes using 35% hydrogen peroxide. High weight reduction up to 95.2% in the presence of copper sulfate and 87.8% in the presence of ferrous sulfate was reported [8, 29, 30]. Recently, synthesis of nano-catalysts of specific size and shape was developed to allow facile movement of materials in the reacting phase and the control over morphology of nanostructures to tailor their physical and chemical properties. Nano-catalyst systems encompassing a paramagnetic core allow rapid and selective chemical transformations with excellent product yield coupled with the ease of catalyst separation and recovery [31].

Talking about catalysis, it is nowadays indispensable to spend some words on the topic biocatalysis; hence, the application of enzymes as highly selective and active catalysts produced by Mother Nature. Not only do enzymes carry out the desired reactions under mild conditions of temperature and pressure, which is analogous to the above-discussed energy efficiency principle. Moreover, they are predestined to drive reactions of renewable materials (analogy to the renewable feedstock principle) and, in some cases, even enable reactions that are not accessible by using traditional catalysts, such as the generation of some enantio-pure products [32]. In addition, biocatalysts in a free or immobilized form are to an increasing extent applied for bioremediation, hence, the mitigation of pollutants from the ecosphere [33, 34].

In the context of catalysis, microwave-driven chemical reaction for organic chemistry is an emerging field to drastically reduce reaction times from days to only seconds. During the last decades, this technique has matured from a laboratory shenanigan to a well-established technology, now already industrially applied [35].

2.10. Design for degradation

One of the most important objectives of green chemistry is maximizing the production with minimizing unwanted by-products. Designing of products and processes that display reduced impact on humans and the environment, such as creating sustainable mortar composite that could be considered as an value-added product suitable for various applications as inert matrix for immobilization of some low and intermediate levels of radioactive wastes, decorative tiles, building bricks, and light concrete, is reported. In this case, highly reactive hydroxyl radicals react with the organic moieties of the spinning fiber wastes either by subtracting ions of hydrogen or by addition to the unsaturated site to yield organic radicals, which are readily oxidized by oxygen. Therefore, the end products of the degradation process were only carbon dioxide and water [8–11].

2.11. Real-time analysis for pollution prevention

With advancements in chemistry, the production of numerous toxic chemicals is a serious problem for the environment. One of the basic concepts of green chemistry is familiar to pollution prevention practitioners. Less hazardous materials in chemical formulations and reducing waste formation have been sounded for many years. Consequently, green chemistry aims at eliminating the usage and generation of hazardous substances by designing better manufacturing processes for chemical materials with minimum waste production by real-time monitoring of running processes. This consequently enables a timely intervention right before waste or toxins are generated [36].

2.12. Inherently safer chemistry for accident prevention

It is of outstanding importance to avoid highly reactive chemicals that could potentially cause accidents during the reaction. Substance and the form of a substance used in a chemical process should be chosen in such a way to minimize the potential for chemical accidents, including toxin releases, explosions, and fire formation. For example, the most abundant solution medium, water, could accidentally cause an explosion by flowing into a tank containing methyl isocyanate gas, releasing a large amount of methyl isocyanate into the surrounding area. Other well-known materials, which undergo reactions of often disastrous outcome with water, are found among alkali metals. If an alternative reaction had been developed that did not use this reagent, the risk of explosion even causing death would have been minimized.

Intrinsically, safe chemistry can also be carried out in flow mode, using tubular microreactors with reaction channels of tiny diameter. Such flow chemistry approaches drastically reduce the reaction volume, the reaction time, and catalyst requirement, intensifies the processes by boosting the space/time yield, opens new process windows in terms of extreme temperature and pressure conditions to be applied, and, moreover, even allows to carry out highly dangerous reactions in a safe way [37, 38]. In addition, the application of flow chemistry in microreactors also displays a strategy to overcome classical drawbacks of microwave-driven processes, such as the restricted penetration depth of microwaves into absorbing media [39, 40].

Author details

Hosam El-Din Mostafa Saleh[1]* and M. Koller[2]

Address all correspondence to: hosamsaleh70@yahoo.com

1 Radioisotope Department, Nuclear Research Center, Atomic Energy Authority, Egypt

2 University of Graz, Office of Research Management and Service, c/o Institute of Chemistry, NAWI Graz, Graz, Austria

References

[1] Anastas PT, Warner JC. Green Chemistry: Theory and Practice. New York: Oxford University Press; 1998

[2] Schnitzer H, Ulgiati S. Less bad is not good enough: Approaching zero emissions techniques and systems. Journal of Cleaner Production. 2007;**15**(13-14):1185-1189

[3] Peters M., von der Assen N. It is better to prevent waste than to treat or clean up waste after it is formed – Or: What Benjamin Franklin has to do with "Green Chemistry" Green Chemistry, **18**(5):1172-1174

[4] Trost BM. The atom economy-a search for synthetic efficiency. Science. 1991:**254**(5037); 1471-1477

[5] Kondo T. On inventing catalytic reactions via ruthena- or rhodacyclic intermediates for atom economy. Synlett. 2008;**2008**(5):629-644

[6] Banert K, Plefka O. Synthesis with perfect atom economy: Generation of Diazo ketones by 1, 3-dipolar Cycloaddition of nitrous oxide at cyclic alkynes under mild conditions. Angewandte Chemie International Edition. 2011;**50**(27):6171-6174

[7] Baghbanzadeh M, Pilger C, Kappe CO. Rapid nickel-catalyzed Suzuki– Miyaura cross-couplings of aryl Carbamates and Sulfamates utilizing microwave heating. The Journal of Organic Chemistry. 2011;**76**(5):1507-1510

[8] Saleh HM, Eskander SB, Fahmy HM. Mortar composite based on wet oxidative degraded cellulosic spinney waste fibers. International journal of Environmental Science and Technology. 2014;**11**(5):1297-1304

[9] Eskander SB, Saleh HM, Fahmy HM. Incorporation of the spinning wastes in cement and mortars. Journal of Radiation Research and Applied Sciences. 2009;**2**(1):119-136

[10] Eskander SB, Saleh HM. Cement mortar-degraded spinney waste composite as a matrix for immobilizing some low and intermediate level radioactive wastes: Consistency under frost attack. Journal of Nuclear Materials. 2012;**420**(1-3):491-496

[11] Saleh HM, Eskander SB. Characterizations of mortar-degraded spinney waste composite nominated as solidifying agent for radwastes due to immersion processes. Journal of Nuclear Materials. 2012;**430**(1-3):106-113

[12] Saleh HM, Tawfik ME, Bayoumi TA. Chemical stability of seven years aged cement–PET composite waste form containing radioactive borate waste simulates. Journal of Nuclear Materials. 2011;**411**(1-3):185-192

[13] Eskander SB, Bayoumi TA, Saleh HM. Performance of aged cement-polymer composite immobilizing borate waste simulates during flooding scenarios. Journal of Nuclear Materials. 2012;**420**(1-3):175-181

[14] https://www.acs.org/content/acs/en/greenchemistry/what-is-green-chemistry/principles/green-chemistry-principle-3.html

[15] Eskander SB, Bayoumi TA, Saleh HM. Leaching behavior of cement-natural clay composite incorporating real spent radioactive liquid scintillator. Progress in Nuclear Energy. 2013;**67**:1-6

[16] Bayoumi TA, Saleh HM, Eskander SB. Solidification of hot real radioactive liquid scintillator waste using cement-clay composite. Monatshefte fur Chemie - Chemical Monthly. 2013;**144**(12):1751-1758

[17] Kerton FM. Alternative Solvents for Green Chemistry. Clark J: RSC Green Chemistry Series, Series Editors; 2009

[18] Yanlong G, Jérôme F. Glycerol as a sustainable solvent for green chemistry. Green Chemistry. 2010;**12**:1127-1138

[19] Koller M, Niebelschütz H, Braunegg G. Strategies for recovery and purification of poly [(R)-3-hydroxyalkanoates](PHA) biopolyesters from surrounding biomass. Engineering in Life Sciences. 2013;**13**(6):549-562

[20] Samorì C, Basaglia M, Casella S, Favaro L, Galletti P, Giorgini L, et al. Dimethyl carbonate and switchable anionic surfactants: Two effective tools for the extraction of polyhydroxyalkanoates from microbial biomass. Green Chemistry. 2015;**17**(2):1047-1056

[21] Koller M, Bona R, Chiellini E, Braunegg G. Extraction of short-chain-length poly-[(R)-hydroxyalkanoates] (scl-PHA) by the "anti-solvent" acetone under elevated temperature and pressure. Biotechnology Letters. 2013;**35**(7):1023-1028

[22] Murugesan S, Iyyaswami R. Nonionic surfactants induced cloud point extraction of Polyhydroxyalkanoate (PHA) from Cupriavidus necator. Separation Science and Technology. 2017;**52**(12):1929-1937 (just-accepted)

[23] Murugan P, Han L, Gan CY, Maurer FH, Sudesh K. A new biological recovery approach for PHA using mealworm, Tenebrio Molitor. Journal of Biotechnology. 2016;**239**:98-105

[24] Shahzad K, Kettl KH, Titz M, Koller M, Schnitzer H, Narodoslawsky M. Comparison of ecological footprint for biobased PHA production from animal residues utilizing different energy resources. Clean Technologies and Environmental Policy. 2013;**15**(3):525-536

[25] https://www.acs.org/content/acs/en/greenchemistry/what-is-green-chemistry/principles/green-chemistry-principle--7.html

[26] Narodoslawsky M. Structural prospects and challenges for bio commodity processes. Food Technology and Biotechnology. 2010;**48**(3):270-275

[27] http://www.orgchemboulder.com/Labs/Handbook/GreenChemistry.pdf

[28] Weitkamp J. Katalyse. Chemie Ingenieur Technik. 2003;**75**(10):1529-1533

[29] Shatta HA, Bayoumi TA, Saleh HM. Catalytic wet oxidative degradation of filter paper waste simulates, isotope and. Radiation Research. 2005;**37**(6):1599-1611

[30] Saleh HM. Treatment and Solidification of Hazardous Organic Wastes: Radioactive Cellulose-Based Wastes, Book ISBN 978-3-659-18564-9. Germany: LAP Lambert Academic Publishing; 2012

[31] Polshettiwar V, Varma RS. Green chemistry by nano-catalysis. Green Chemistry. 2010; **12**:743-754

[32] Schmid A, Dordick JS, Hauer B, Kiener A. Industrial biocatalysis today and tomorrow. Nature. 2001;**409**(6817):258

[33] Mohan T, Rathner R, Reishofer D, Koller M, Elschner T, Spirk S, Heinze T, Stana-Kleinschek K, Kargl R. Designing hydrophobically modified polysaccharide derivatives for highly efficient enzyme immobilization. Biomacromolecules. 2015;**16**(8):2403-2411

[34] Rathner R, Petz S, Tasnádi G, Koller M, Ribitsch V. Monitoring the kinetics of biocatalytic removal of the endocrine disrupting compound 17α-ethinylestradiol from differently polluted wastewater bodies. Journal of Environmental Chemical Engineering. 2017;**5**(2):1920-1926

[35] Kappe CO. Microwave dielectric heating in synthetic organic chemistry. Chemical Society Reviews. 2008;**37**(6):1127-1139

[36] Pollution Prevention by Utilizing Green Chemistry, Office of Compliance Assistance & Pollution Prevention, Fact sheet. 2006. p. 106

[37] Hessel V. Novel process windows–gate to maximizing process intensification via flow chemistry. Chemical Engineering & Technology. 2009;**32**(11):1655-1681

[38] Razzaq T, Glasnov TN, Kappe CO. Continuous-flow microreactor chemistry under high-temperature/pressure conditions. European Journal of Organic Chemistry. 2009;**2009**(9): 1321-1325

[39] Glasnov TN, Kappe CO. Microwave-assisted synthesis under continuous-flow conditions. Macromolecular Rapid Communications. 2007;**28**(4):395-410

[40] Damm M, Glasnov TN, Kappe CO. Translating high-temperature microwave chemistry to scalable continuous flow processes. Organic Process Research & Development. 2009;**14**(1):215-224

Section 2

Green Separation Techniques

Chapter 2

Green Separation of Bioactive Natural Products Using Liquefied Mixture of Solids

Siti Zullaikah, Orchidea Rachmaniah,
Adi Tjipto Utomo, Helda Niawanti and Yi Hsu Ju

Additional information is available at the end of the chapter

http://dx.doi.org/10.5772/intechopen.71755

Abstract

Bioactive natural products are secondary metabolites of plants and animals generated through various biological pathways. They are the main sources of new drugs, functional food and food additives. Since their contents in plant and animal tissues are extremely small compared to those of primary metabolites, the separations of bioactive principles from complex matrixes are often the inherent bottleneck in the utilization of bioactive natural products. A novel separation technique based on a liquefied mixture of solids at its eutectic compositions is presented in this chapter. The mixture can be prepared from natural primary metabolites and therefore can be considered as a green solvent. The separation of bioactive compounds (γ-oryzanol) from rice bran oil-based biodiesel using green methods with minimum energy requirement is discussed. Other applications for separations of alkaloid and phenolic compounds from their plant matrices are also presented. Different raw materials require different separation techniques due to the presence of different impurities, and the current trend is to use green methods with minimum energy requirement. This overview of recent technological advances, discussion of pertinent problems and prospect of current methodologies in the separation of bioactive natural products may provide a driving force for the development of novel separation techniques.

Keywords: green separation, bioactive natural products, deep eutectic solvent, natural deep eutectic solvent

1. Introduction

Bioactive compounds of plants, also known as natural products, are produced as secondary metabolites. Unlike primary metabolites, such as carbohydrates, proteins, fats, amino acids, nucleic acids and organic acids, which are essential to perform the metabolic rules involved in the life process, they have no apparent direct functions in growth, development and reproduction. They are often differentially distributed among limited groups of plants and only present in very low quantities in plants. Though, in principle, they are inessential to life, many secondary metabolites found in plants have roles in defense against predators (herbivores, pests and pathogens), competition and facilitating the reproduction process. However, many of them still remain unknown in their functions. Previously, secondary metabolites were generally thought to be waste products of plants without apparent function. Nowadays, they represent an important source of biological active compounds which are very important for the development of food and pharmaceutical industries.

Since their content is small and different raw materials require different isolation techniques due to the presence of different impurities, the extractions of bioactive principles from complex matrixes are often the inherent bottleneck in the utilization of bioactive natural products. The extraction techniques can be classified into conventional and modern ones [1]. The conventional techniques include maceration, percolation, Soxhlet extraction and solvent extraction. They are typically characterized by long extraction time, high cost due to the requirement of large volume of solvents, low yield and the use of toxic and flammable solvents. The modern techniques include enzyme-assisted extraction, ultrasound-assisted extraction, microwave-assisted extraction, subcritical and supercritical fluid extraction and high pressure-assisted extraction. They generally have shorter extraction time, lower cost, higher purity of the extracted compounds and much better efficiency [2]. However, there are still issues associated with conventional and modern extraction techniques, including the toxicity of solvent, thermal instability, solubility and poor selectivity. In addition, the type and concentration of solvents, their moisture contents, recovery of bioactive compounds and changes of bioactive compounds during extraction due to ionization, hydrolysis, esterification and oxidation need to be considered [3]. Water can be used as solvent in the extraction of bioactive compounds. However, water is only effective to extract polar and hydrophilic bioactive compounds but less effective to extract non-polar and hydrophobic ones. In addition, impurities extracted by water become another problem for further purification steps.

Green solvents have been explored to replace traditional hazardous solvents that are used extensively in industry. To be qualified as a green solvent, a solvent should be non-toxic, non-volatile, biodegradable without generating any toxic and persistent metabolite, inflammable, recyclable, relatively cheap and available in a large quantity [4–6]. Water at sub- and supercritical conditions and CO_2 at supercritical condition are examples of green solvents that have been used in industrial scale.

Recently, ionic liquids (ILs) have been developed as green solvents. Ionic liquids are molten salts, mixtures of bulky and asymmetric organic cations and organic or inorganic anions, with melting points usually below 100°C [2, 7]. They have some attractive attributes, such as non-flammable, high thermal and chemical stabilities and low vapor pressure [1, 4, 8–10]. The combination of different cations and anions makes ILs have a tunable nature, i.e. polarity and other properties with their own unique structures and properties. Therefore, there has been an increase in the use of ILs as green solvent for the extraction, separation and purification of natural bioactive compounds [2]. However, there are concerns about the application of ILs related to the toxicity of these compounds, their potential effects on health and the environment and the high cost associated with their synthesis and purification requirements [8, 11].

To overcome the drawbacks of ILs, deep eutectic solvents (DESs) have been developed. They have physicochemical properties similar to those of ILs. In addition, DESs are biodegradable, less toxic and cheaper than ILs [12]. DESs are formed from mixtures of two or more Lewis acids and bases or Bronsted-Lowry acids and bases that have the lowest freezing points compare to their starting constituents [13]. The physical structures of some DESs are similar to those of ILs. However, DESs in general are different in terms of the source of the starting ingredients and the chemical formation process.

2. Eutectic solvents

Deep eutectic solvent (DES) is a eutectic mixture of two or more compounds which has a melting point much lower than either of the individual components [6, 14, 15]. A eutectic mixture is the condition when the molar ratio of the component gives the lowest melting point as represented in **Figure 1**. DES was first introduced by Abbott et al. [14] who studied the properties of choline chloride (ChCl)/urea mixture. Both ChCl and urea have melting points of 302 and 133°C, respectively. However, at the eutectic composition (1:2 ChCl/urea molar ratio), the mixture melts at 12°C making it liquid at room temperature.

Typically, DESs are mixtures of quaternary ammonium halide salts and hydrogen bond donors (HBDs). Various quaternary ammonium halide salts and HBDs which can form DESs are shown in **Figure 2**. One of the most widely used ammonium quaternary salt for DESs

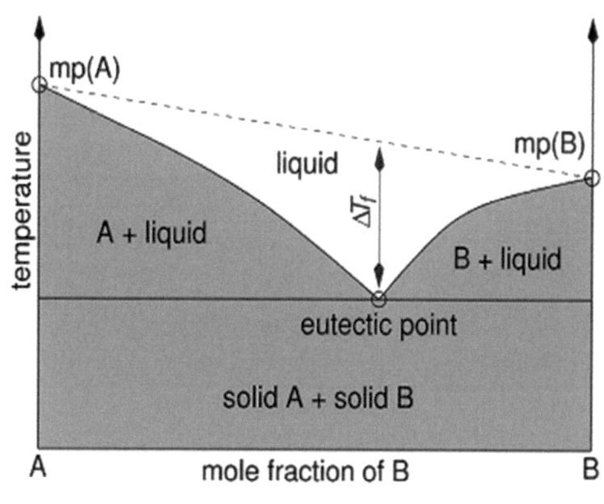

Figure 1. Schematic representation of a eutectic point on a two-component phase diagram [15].

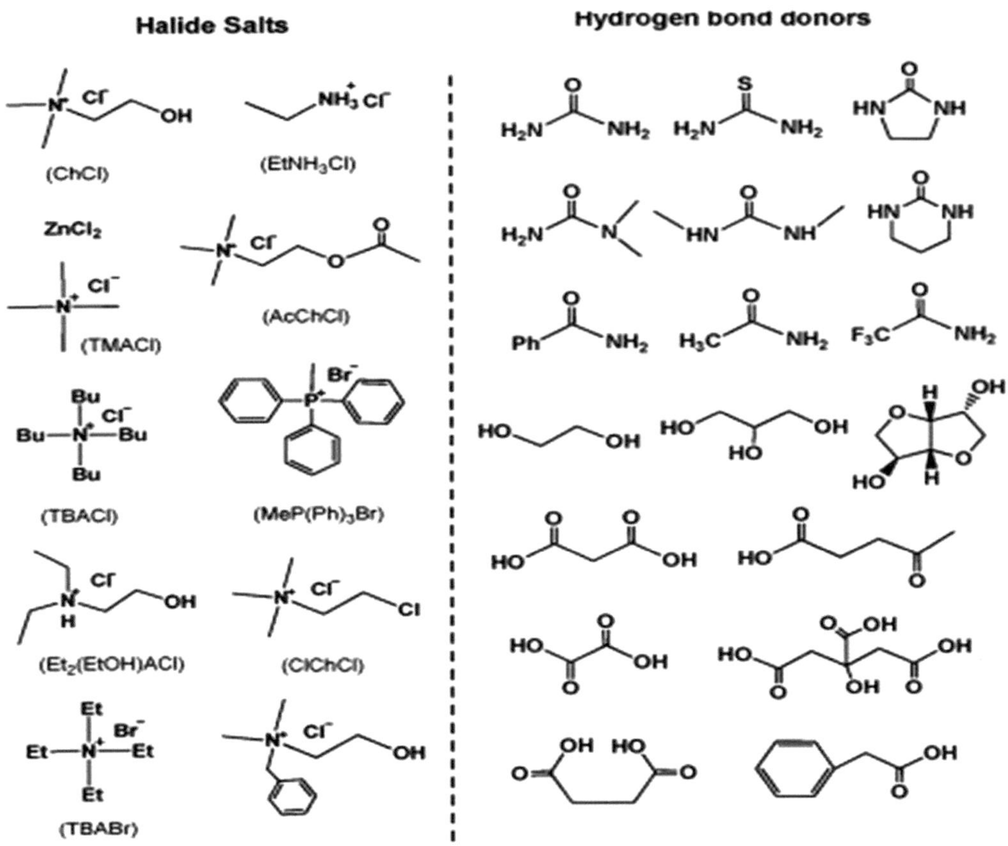

Figure 2. Halide salts and HBD for DES [13].

formation is ChCl. ChCl is cheap, biodegradable, non-toxic and can be easily extracted from biomass or synthesized from fossil fuel, while the HBD could be amides (e.g. urea) [14], carboxylic acids (e.g. oxalic acid), alcohols (e.g. glycerol) [13, 16], sugars or sugar analogues, amino or organic acids and alkylsulfates or alkyl phosphates.

The main advantage of DESs over the previous generation of ILs is that they are easier and simpler to make. The solids are mixed in gentle heat until they melt, and when they cool down, they remain in the liquid form. No purification steps are required since there is no formation of new salt and the final purity is determined by the purities of the starting materials. The design of DESs is simpler and more flexible since no reaction takes place, and therefore it does not have any strict stoichiometry limitation. The interaction between HBD and hydrogen bond acceptor (HBA) will form a liquid in their relative molar composition. The behaviors and properties of DESs can be tuned by varying the HBD and its molar ratio in the mixture [13]. DESs have moderate polarity, stability and distributed negative charge like ILs, but they are biodegradable, readily available and less toxic since DESs can be based on bulk natural product such as carbohydrates (fructose, glucose, mannose, maltose and α-cyclodextrin), sugar alcohols (sorbitol) or citric acids with urea (or N,N-dimethylurea) and inorganic salts (NH_4Cl and $CaCl_2$) [10, 17].

Another term of DESs was introduced by Gutiérrez et al. [16] as low transition temperature mixtures (LTTMs). They are the right combinations of different molar ratios between HBD and HBA, such as lactic acid/alanine = 9:1, lactic acid/ChCl = 2:1, lactic acid/histidine = 9:1, etc. The formed liquid mixtures have glass transitions instead of melting points [18]. A complete characterization of physical properties (density, viscosity, surface tension, glass transition temperature) of LTTMs, i.e. lactic acid/ChCl = 2:1, was reported by Francisco et al. [18].

Recently, Choi et al. [19] have first reported a large number of DESs which were mixtures of ChCl with any primary metabolites, e.g. sugars, sugar alcohols, organic acids and amino acids. This type of DESs is termed as natural deep eutectic solvents (NADESs). NADESs derived from major compounds always present in all microbial, mammalian and plant cells form liquid crystals. Therefore, NADESs are believed as the third liquid phase present in the cells, in addition to the already considered known phases, i.e. water and lipid [membranes] are responsible for transporting of numerous compounds with intermediate polarity in high concentration that neither dissolve in the lipid nor in the aqueous phase. Rutin, a flavonoid which is barely soluble in water, has a solubility of 0.28 mmole/mL in glucose-choline chloride-water (GCH, molar ratio = 2:5:5), thus 50–100 times higher than that in water [20], whereas, paclitaxel and ginkgolide B, which are completely water-insoluble compounds, have solubilities of 0.81 and 5.85 mg/mL, respectively, with the same NADES type [19]. In the case of macro molecules, such as DNA and starch, they also show higher solubilities in NADESs than those in water, i.e. 1.20 and 7.55 g/mL, respectively. These facts are in line with the hypothesis of the existing alternative liquid phase to water in nature of poorly water-soluble molecules including high molecular weight molecules. Though GCH (2:5:5) has polarity close to water, GCH as a NADES shows different performance than water. It indicates that NADESs have huge potential for many practical applications since they can be designed as tailor-made solvents.

3. Extraction of bioactive compounds from rice bran oil (RBO)-based biodiesel by choline chloride-based deep eutectic solvent

Rice bran is a promising raw material for biodiesel production. It is relatively cheap, abundant and traditionally used as cattle food. The annual worldwide production of rice bran oil (RBO) could reach 8 million tons if all rice bran produced is harnessed for oil production [21]. RBO is rich in naturally occurring biologically active and antioxidant compounds, such as phytosterols, γ-oryzanol, tocopherols and tocotrienols (tocols) [22]; however, the refinery of crude RBO requires extra processing steps due to high concentrations of free fatty acids (FFA), unsaponifiable matter and dark color [21] making it uncompetitive against other edible cooking oils, such as palm oil, soybean oil and rapeseed oil.

Since RBO contains high FFA, conventional process to produce biodiesel using base catalyst is unsuitable to convert RBO into biodiesel (fatty acid methyl esters, FAME) due to the formation of soap. Several methods have been proposed to convert crude RBO into FAME, i.e. a two-step acid-catalyzed process [23], a three-step method using both acid and base catalysts [24] and a supercritical methanol method [25]. More recently, in situ process to produce biodiesel from rice bran without any pretreatment has also been proposed by Zullaikah et al. [26].

Crude biodiesel produced from RBO using acid-catalyzed methanolysis method typically contains about 89% FAME, 4% triglycerides (TG), 4% diglycerides (DG), 0.3% monoglycerides (MG), 0.05% FFA and 2.55% bioactive compounds, mainly phytosterols and γ-oryzanol. To meet biodiesel standard as fuel (such as [27]), purification process is required to increase FAME content to at least 96.5% and decrease unreacted oil (TG, DG, MG and FFA) contents. Crude biodiesel produced from RBO through acid-catalyzed methanolysis method has different impurity compositions to those produced from edible oil (palm oil, soybean oil and rapeseed oil) using base catalyst, and therefore a different purification process is required. Besides that, since crude biodiesel from RBO contains bioactive compounds, a purification process which is able to capture those bioactive compounds will be of interest. According to Ju and Zullaikah [22], bioactive compounds were not much degraded during acid-catalyzed methanolysis. These bioactive compounds could subsequently be isolated and sold separately as high-value by-products and therefore could reduce the production cost of biodiesel.

DES is a promising solvent to be employed in the purification of crude biodiesel since it is inflammable, non-toxic, biodegradable and considered as a green solvent [12]. One of the most commonly used DESs for biodiesel purification process is a mixture of choline chloride (an ammonium quaternary salt) and ethylene glycol as HBD at a molar ratio of 1:2. The mixture of ethylene glycol and ChCl is called ethaline, and some physical properties of ethaline were shown in **Table 1**. Ethaline has two interactions. The first is between the chlorine anion and the hydroxyl hydrogen atom of choline, and the second is between the anion and the hydroxyl hydrogen atoms in ethylene glycol [28]. Ethaline has a strong interaction with unreacted oil (DG, MG and FFA) and unsaponifiable matter (bioactive compounds) from crude RBO-based biodiesel due to the presence of hydroxyl groups on those compounds. On the other hand, the interaction between ethaline and FAME is relatively weak since FAME has no hydroxyl group.

Based on the previous research [29], ethylene glycol interacted into each other by making hydrogen bonding in cyclical pattern and the distance of H-O bond was 1.944 Å, whereas, ChCl in crystalline structure has three bonds consisted of C-N, C-O and C-C, and the distance of each bond is 0.01 Å. ChCl is difficult to convert into liquid at room temperature due to the small distance of ChCl bonds. Based on Wagle et al. [29], DES from ChCl and ethylene glycol has two interactions of C-H-O. The first one was between oxygen from ethylene glycol and methyl proton from ChCl, and the distance was 2.146–2.440Å. The

Melting point (K)[a]	207.15
Viscosity (cP)[b]	36
Conductivity (mS/cm)[b]	7.61
Density (g/cm)[b]	1.12
Surface tension (mN/m)[b]	49

[a][30].
[b][15].

Table 1. Physical properties of ethaline (molar ratio of ChCl/ethylene glycol = 1:2).

second one was between hydrogen in ethylene glycol and Cl^- in ChCl, and the distance was 2.271–2.474 Å. Cl^- as anion in ChCl forms a centerpiece by interacting with five hydroxyl groups, one hydroxyl group of choline cation and four hydroxyl groups from two ethylene glycol molecules.

As a novel green solvent, a combination of mechanisms and molecular structure of DES is still unknown [31]. Therefore, FT-IR analysis was conducted in this study to determine the functional groups of DES, and the results are shown in **Figure 3** and **Table 2**. The peak in the region of 3200–3500 showed the presence of O-H groups in choline chloride and ethylene glycol-based DES which is in agreement with that reported by Aissaoui [32]. **Figure 3** describes that there is a shift of the O-H stretching band in the DES compared to that in the choline chloride and ethylene glycol. This shift is due to the electrons of oxygen that are transferred to the hydrogen bonds making the constant force lower and resulting in a change in the vibrational state. The shifting of O-H stretching vibration indicates the presence of hydrogen bonds in DES [32]. The peaks in the region of 3000-2800 on choline chloride, ethylene glycol and DES show the presence of C-H, CH_3 and CH_2 stretching bands. Meanwhile, the N-H stretching bands overlap with the C-H vibrational bands in the region of 3000-2800 cm^{-1} [33]. As shown

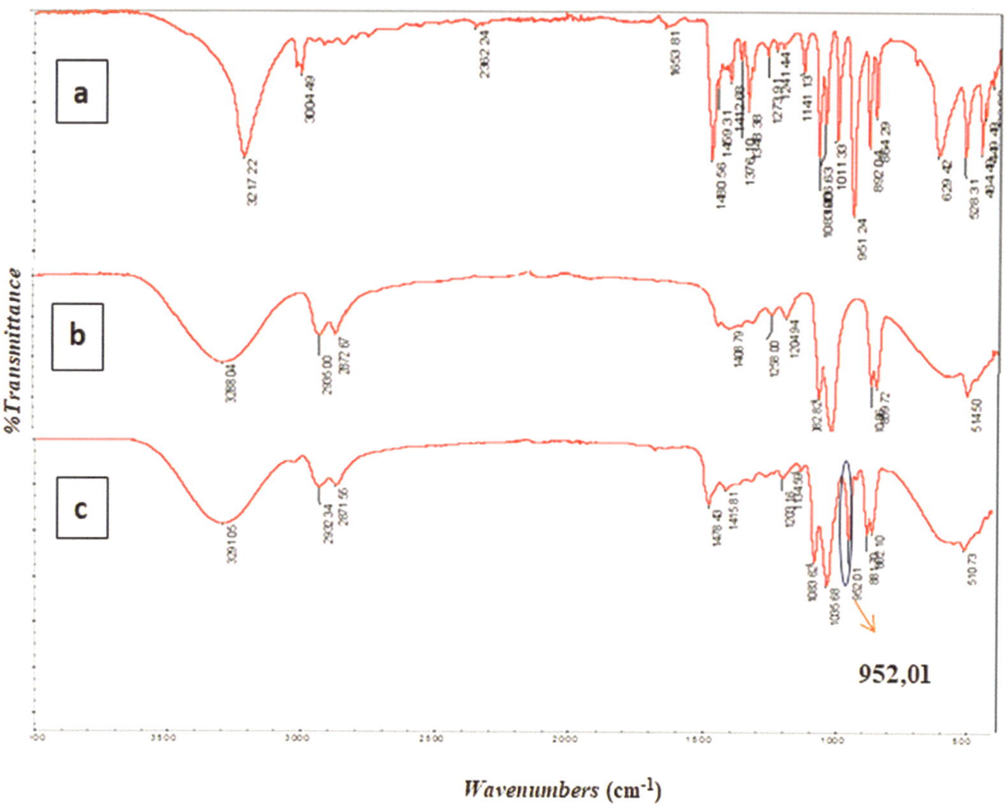

Figure 3. FT-IR analysis: (a) choline chloride, (b) ethylene glycol, (c) ethaline (molar ratio of choline chloride/ethylene glycol = 1:2).

Wavenumber	Functional group
3200–3500	O-H (alcohol)
2845–3000	C-H *stretching*, CH_2 *stretching*, CH_3 *stretching*
496–700	C-X stretching (X = F, Br, Cl or I)
1210–1150	Tertiary amine (C-N)
1000–1350	C-C *stretch*
[a][33].	

Table 2. Wavenumber and functional group of ethaline analysis by FT-IR[a].

in **Figure 3**, the stretching vibration at 2500-3100 regions in choline chloride is invisible after the formation of DES. The presence of Cl^- in DES is shown at 600 and 408 cm^{-1}. **Figure 3** also describes that the DES had a vibration pattern similar to ethylene glycol as HBD, except that the peak is at 952.01 cm^{-1}. The peak appears on DES is in the region of 935-955 cm^{-1} indicating the identity of ammonium structure of DES [32]. The FT-IR analysis describes that the establishment of DES does not lead to the formation of new functional groups in the mixture.

The purification process of crude RBO by using DES is relatively simple and can be described as follows. Crude RBO-based biodiesel and DES were mixed in a stopper glass (50 mL) at a certain molar ratio (1:8). The mixture of biodiesel and DES was heated at a certain temperature (30°C) under stirring at 300 rpm. Afterwards, the mixture was let to settle for 2 h at ambient temperature so that two layers were formed. The upper layer was biodiesel (FAME)-rich phase and the bottom layer was DES-rich phase containing biodiesel impurities including bioactive compounds. The biodiesel-rich phase (upper layer) was then separated from DES-rich phase by using a separator funnel. There are several factors that influence the purification process, such as extraction time, extraction temperature and molar ratio of DES/RBO-based biodiesel. However, this chapter only discusses the effect of extraction time on FAME recovery, removal of unreacted oil and bioactive compounds and the final biodiesel composition.

Figure 4 shows the effects of extraction time on the biodiesel recovery and the removal of unreacted oil and bioactive compounds. Extraction time is one parameter that influences liquid–liquid extraction. The unreacted oil and bioactive compounds diffuse from biodiesel-rich phase to DES-rich phase. More unreacted oil and bioactive compounds migrate to DES-rich phase with longer contact time between crude biodiesel and DES. FAME recovery {(FAME in product/FAME in sample)×100%} also increased from 63.38 to 87.31% as the extraction time was extended from 15 to 240 min as shown in **Figure 4**.

The removal efficiency of unreacted oil and bioactive compounds has similar trend with time as shown in **Figure 4**. However, the removal efficiency of each compound is different. Since TG have no OH^- group, their removal efficiency was lower than the other compounds with OH^- groups, such as DG, MG, FFA and bioactive compounds. TG removal efficiency was practically unaffected by extraction time. TG removal at extraction times of 15 and 240 min were 41.32 and 39.69%, respectively. However, the removal efficiencies of unreacted oil (DG, MG and FFA) and bioactive compounds increased with time. DG removal increased from

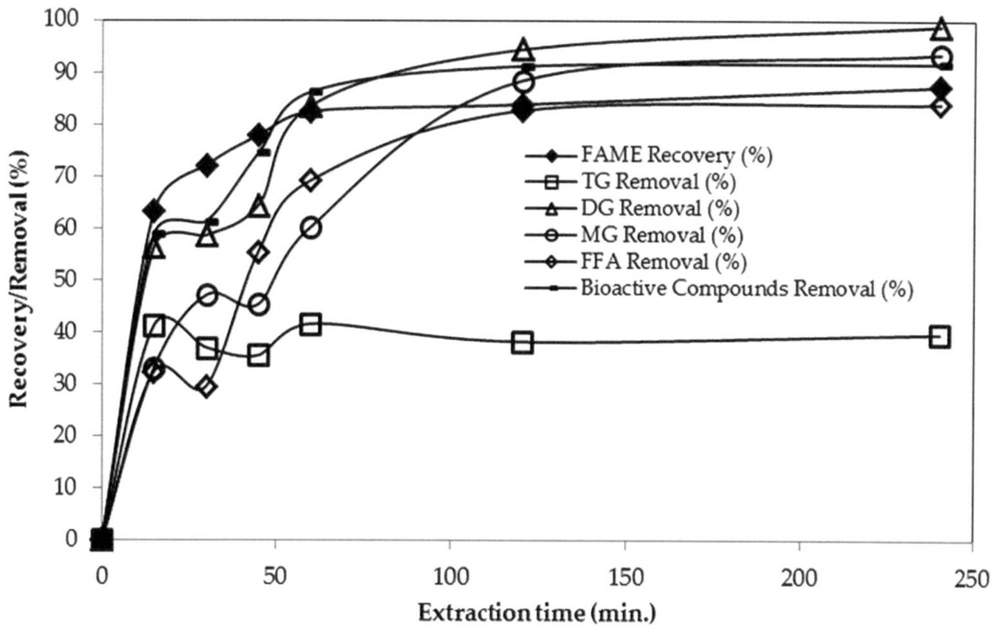

Figure 4. Effect of extraction time on recovery of FAME, unreacted oil and bioactive compound removal. Operation conditions, T = 30°C; molar ratio of ChCl, ethylene glycol = 1:2; and molar ratio of RBO-based biodiesel, DES = 1:8.

56.42 to 98.88% as the extraction time was increased from 15 to 240 min, while that of MG increased from 33.15 to 93.52%. The removal efficiency of FFA was lower than DG and MG, even though they have OH- group. This is probably because FFA content in RBO-based biodiesel was much lower than those of DG and MG. Since bioactive compounds in RBO, such as γ-oryzanol, phytosterols and tocols, have OH- group, they can make hydrogen bonding with DES, and their removal efficiencies were high. The removal of bioactive compounds increased from 58.91 to 91.70% with the increasing extraction time from 15 to 240 min.

The effect of extraction time on the contents of FAME, unreacted oil and bioactive compounds is shown in **Figures 5** and **6**, respectively. FAME content increased with extraction time (**Figure 5**), while those of unreacted oil and bioactive compound content decreased with extraction time (**Figure 6**). Longer extraction time provides longer contact time between DES- and RBO-based biodiesel. Therefore, more unreacted oil and bioactive compounds diffuse to DES through the formation of hydrogen bond. Since unreacted oil and bioactive compounds were removed from RBO-based biodiesel, the FAME content increased. The FAME content after 240 min purification using DES was higher than 96.5%, exceeding that specified by the European biodiesel standard [27]. FFA is undesirable in biodiesel since it causes negative impacts, such as less oxidation stability and corrosion of vital engine components. The FFA content in biodiesel is characterized by acid value. The acid value of RBO-based biodiesel produced using acid-catalyzed methanolysis method was 0.098 mg KOH/g. This acid value was already lower than the maximal acid value stated in several biodiesel standards such as EN 14214 [27] (0.5 mg KOH/g). FFA content in biodiesel after purification was decreased from 0.05 to 0.01% (**Figure 6**). This showed that the removal efficiency of FFA using DES as

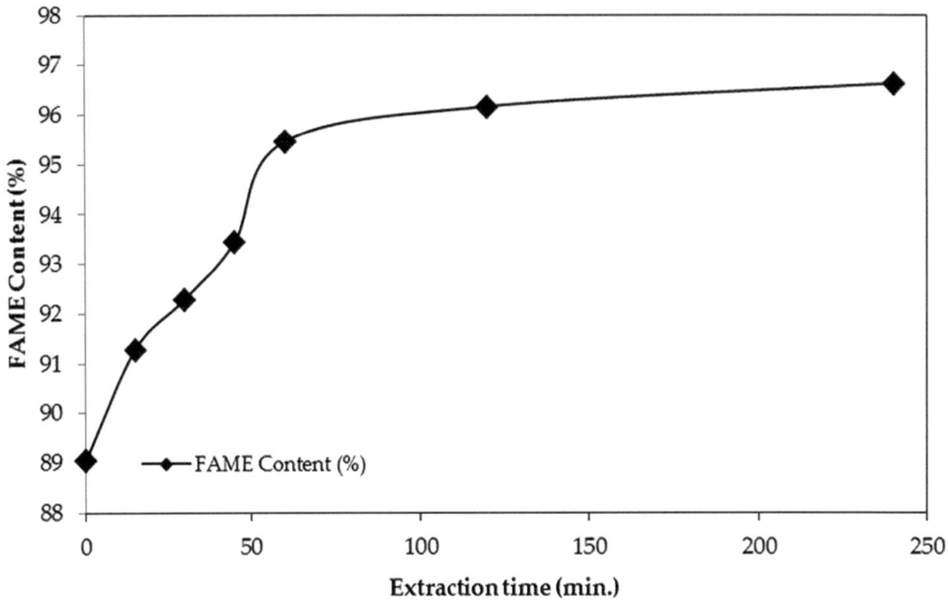

Figure 5. Effect of extraction time on FAME content. Operation conditions, T = 30°C; molar ratio of ChCl, ethylene glycol = 1:2; and molar ratio of RBO-based biodiesel, DES = 1:8.

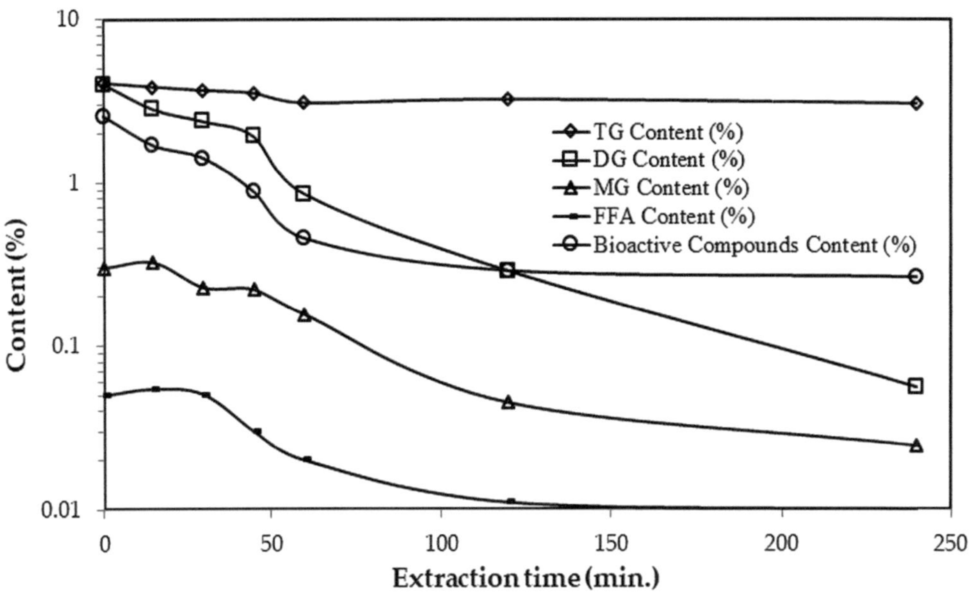

Figure 6. Effect of extraction time on unreacted oil and bioactive compound content. Operation conditions, T = 30°C; molar ratio of ChCl, ethylene glycol = 1:2; and molar ratio of RBO-based biodiesel, DES = 1:8.

Property	Unit	Experimental value	EN 14214 [27]
Ester content	% (m/m)	96.65	Min 96.5
Acid value	mg KOH/g	0.02	Max 0.5
MG content	% (m/m)	0.02	Max 0.8
DG content	% (m/m)	0.06	Max 0.2
TG content	% (m/m)	3.02	Max 0.2

[a]Operation condition, T = 30°C; molar ratio of ChCl, ethylene glycol = 1:2; molar ratio of RBO-based biodiesel, DES = 1:8, 240 min extraction time.

Table 3. Composition of biodiesel after purification using DES[a].

extraction solvent was high. This case shows that the extraction time can affect the extraction process of FFA in biodiesel.

The composition of biodiesel after purification using DES is compared to that specified by biodiesel standard, EN 1424 [27], as shown in **Table 3**. The ester content, acid value, MG content and DG content meet the values specified by EN 14214 [27]. However, TG content was much higher than those specified by EN 14214 [27]. This is because TG does not have OH^- group so it does not interact with DES used in this experiment. This can be overcome by using a multiple-step separation technique.

According Niawanti and Zullaikah [34], γ-oryzanol in the upper layer (biodiesel-rich phase) was decreased with extraction time from about 4% initially to about 1% after 240 min of extraction. The lowest γ-oryzanol content of 1.18% was obtained after 240 min of extraction time. The reason of this phenomenon is that more γ-oryzanol move from RBO-based biodiesel to DES with longer extraction time. The molar ratio of biodiesel to DES also influences the removal efficiency of γ-oryzanol. The higher molar ratio of DES to RBO-based biodiesel leads to higher removal efficiency of γ-oryzanol since more γ-oryzanol molecules are bound to DES molecules through hydrogen bonding.

4. Extraction of phenolic and alkaloid compounds using natural deep eutectic solvent

In recent years, many herbs and natural compounds have increasingly been receiving public interest as complementary and alternative medicines. The natural product curcumin 1,7-bis(4-hydroxy-3-methoxy phenyl)-1,6-heptadione-3,5-dione is a dietary phytochemical obtained from the dried rhizomes of the turmeric plants. It is a natural bioactive compound that has demonstrated both antioxidant and therapeutic anticancer capabilities. However, it is not yet fully used clinically due to its inherent limitations, i.e. sparing solubility in water and low bioavailability. Curcumin (C) is extracted from *Curcuma Sp.*, i.e. *Curcuma longa,*

Curcuma zedoaria and *Curcuma manga*, a plant of the ginger family, with other two curcuminoid compounds: demethoxycurcumin (DMC) and bisdemethoxycurcumin (BDMC).

Traditionally curcumin has been used as food coloring, flavoring and preservative. Because of its wide spectrum of biological activity, an extensive number of studies have been focused on curcumin. Recently, curcumin has also been shown to display antioxidant, anticancer, antiviral, anti-infectious and anti-amyloidogenic properties. Numerous methods to isolate curcumin as well as other curcuminoids from *C. longa* rhizomes have been reported, such as conventional solvent extraction, hot and cold percolation, the use of alkaline solution and insoluble salt, supercritical carbon dioxide extraction, microwave-assisted extraction and ultrasonic-assisted extraction techniques. However, only a few of them use either green solvent or green process such as pressurized hot water extraction [35]. In the contrary, volatile organic solvents, e.g. methanol, ethanol, acetone and hexane, are still widely applied.

The capabilities of NADES to extract and stabilize bioactive compounds have been investigated by several authors. The solubilities of secondary metabolites such as rutin, quercetin, cinnamic acid, carthamin, taxol and ginkgolide B in different types of NADES have been studied by Choi et al. [19] and Dai et al. [20], while the stabilization ability of NADES on unstable natural colorants, carthamin and anthocyanins, during heating, storage and exposure to light has been reported by Dai et al. [36]. In addition, Bajkacz and Adamek et al. [37] also showed that NADES can be applied for isoflavone extraction. NADES can also be applied for protein stabilization [38], bioavailability improvement [39], antimicrobial agent [40] and bioactivity enhancement of plant extract [41]. Most of them used choline chloride-based NADES, the best combination of NADES which is suitable with their studied compounds. The broad range utilization of NADES showed that NADES leads a novel application in food and pharmaceutical industry.

The selection of solvent for extraction, i.e. liquid-liquid extraction (LLE), depends on its physical properties such viscosity, density and miscibility. It is convenient to select solvent with low viscosity to facilitate mixing as well as maximizing solvent penetration to the plant matrix but with a large density difference for the separation process. The inherent viscous properties of NADESs differ enormously according to their composition, but in all cases, it can be reduced by the addition of a certain amount of water. It should be noted that the addition of water changes the properties of NADESs, i.e. polarity, density and solubilizing and stabilizing capability. However, excessive dilution of NADESs, ca. approximately >50% weight of water, disrupts the special structure of NADESs due to the loss of the existing hydrogen bonds [20]. The viscosities of NADESs can also be decreased by increasing temperature. Generally, NADESs composed of sugar are the most viscous, while choline chloride-based NADESs are less viscous, while the glycerol-based NADESs are the least. Common efforts to minimize the resistance of viscosity and improving the extraction rate, such as mechanical agitation, microwave and ultrasound-assisted extraction, can be used with NADES.

To figure out the broad application of NADES in natural product extraction, this chapter documents the application of NADES on the extraction of curcumin, a low solubility phenolic compound in water, and galantamine, an alkaloid of acetylcholine inhibitor.

Curcuma zedoaria that contains 1.96% ± 0.07% of dry weight of curcuminoids was kindly donated by Sari Herbal (Sukun, Malang, Indonesia). It was chosen in this study since it is traditionally believed as an anticancer medicine in Indonesia. Twelve NADESs were used as solvents for extracting curcuminoids from fine powder of *C. zedoaria*, representing different types of NADESs: (1) ionic type, consisted of organic acids, i.e. citric acid, malic acid and lactic acid, and basic compounds such choline chloride or betaine; (2) neutral type, no ionic constituent, mixed of polyalcohol, i.e. glycerol, glycine, 1–2-propanediol and sugars; (3) acidic type, consisted of neutral compounds such sugars and acidic compounds; (4) basics type, consisted of basic compounds; and the last is (5) amphoteric type, consisted of combination of amino acids and sugars, polyalcohol or acidic compounds. In the case of ionic type of NADESs, it is represented by CCCA-H_2O and CCMA-H_2O; FS-H_2O and FG-H_2O are neutral type and CAS-H_2O and MAS-H_2O for the acidic type, whereas CCGo-H_2O, CCG-H_2O and CCF-H_2O are the basic type. Meanwhile, the amphoteric types were excluded in this study.

A simple extraction protocol was developed to test the capability of NADES to solubilize curcuminoids from plant matrix. The powder of *C. zedoaria* was mixed with NADES in a bottle with cap (powder/NADES ratio = 20 mg powder/3 mL NADES) and stirred (350 rpm) at 40°C for 24 h. Triplicate samples of the resulting solution were diluted with water and analyzed with HPLC-DAD at a wavelength of 421.4 nm. The NADESs were prepared according to Dai et al. [20] with slight modification, i.e. by using freeze-drying instead of vacuum evaporation. The liquefied solid mixture can be called as NADES, when after the freeze-drying process it remains liquid and is visually clear and transparent with no precipitation and crystallization that are formed. It can be kept until a year without any changes in appearances and physical properties, i.e. density and viscosity. In addition, the purity of the individual component of NADES does not affect the NADES properties. As shown in **Figure 7**, the FT-IR spectra of CAS-H_2O prepared with different grades (purity) of citric acid are similar indicating that there are no structural changes.

In analogy with DES, the two components of NADES are particularly bonded by hydrogen bond [19]. The cross-correlation between sucrose and malic acid was observed by ^1H-^1H-nuclear Overhauser enhancement spectroscopy (NOESY). It revealed molecular interactions of protons on the C2 and C3 positions of malic acid with those on the C1 and C2′ of sucrose [19]. This analysis also suggests that water might also participate in the formation of supermolecular structure of NADES [19, 20].

In the case of acidic type of NADESs, e.g. CAS-H_2O and MAS-H_2O, a yellowish color will be observed right after the extraction process takes place. It will be deepened along the elapsed extraction time due to the caramelization of sucrose promoted by the presence of acid. The pH of both CAS-H_2O and MAS-H_2O is two (measured at 10x dilution with Aquadest). The caramelization reaction is the hydrolysis of sucrose by protonation of the glycosidic linkage [42]. Though the extraction process was only heated at 40°C, the effect of acidic condition is comparable with the effect of heating at high temperature at neutral pH [43]. However, neither the density nor viscosity of CAS-H_2O and MAS-H_2O was significantly affected by the hydrolysis reaction (**Figure 8**). Nevertheless, sugar hydrolysis was not observed in other NADESs that consist of sucrose such as in the neutral NADES type, FS-H_2O, although sucrose hydrolysis also occurs at very concentrated sucrose solution even at neutral pH [42].

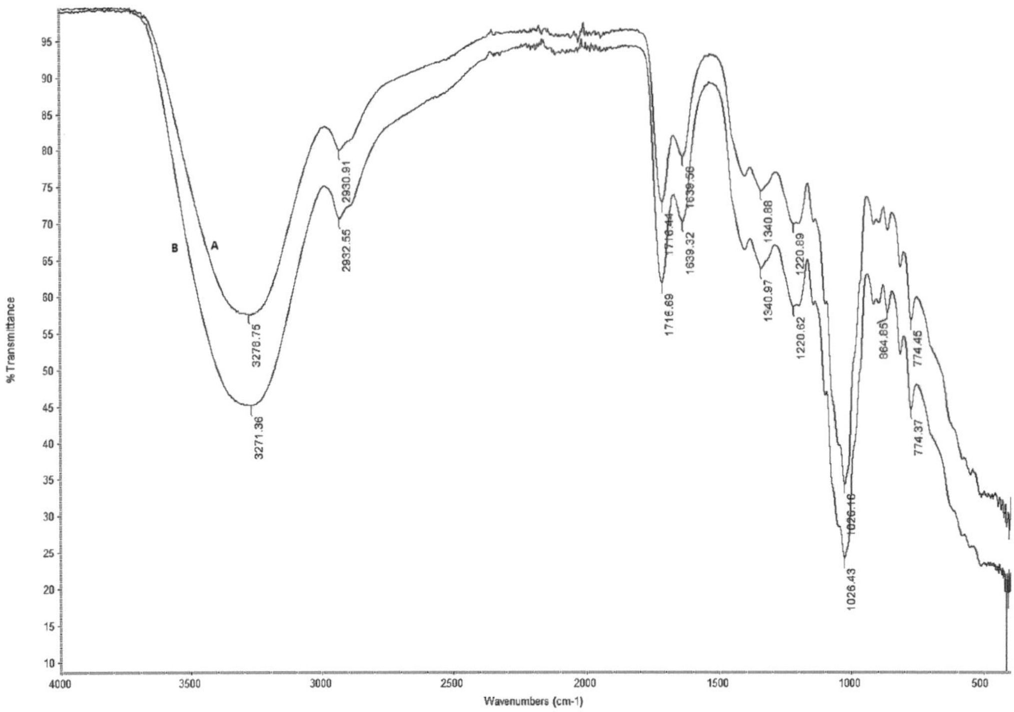

Figure 7. FT-IR spectra of CAS-H_2O, (A) using technical grade and (B) using pure grade of citric acid.

The resulted data of curcuminoid extraction with NADESs are shown in **Figure 9**. It is surprising that in overall NADESs show better extracting capability of curcuminoids than organic solvent, i.e. ethanol, and water although NADESs themselves are water-based solvents (**Figure 9**). Curcuminoids can be extracted by NADES due to the hydrogen bond formed between curcuminoids with NADES suggesting the presence of hydroxyl groups in curcumin, demethoxycurcumin as well as bisdemethoxycurcumin. At the same time, none of curcuminoids was extracted with both ethanol and water using the same extraction protocol as NADES. Curcumin is practically insoluble in water, i.e. ca. 4 ppb (4 μg/L at pH = 7,3) [44]. Curcuminoids were the best extracted by CCMA-H_2O (1:1:3), 0.355 ± 0.019 mg/g, which is in agreement with that reported by Euterpio et al. [35] and Kwon and Chung [45]. A 0.136 mg curcuminoids/g of *C. longa* was obtained using a subcritical mixture of MeOH-H_2O (50:50, v/v) at 135°C, 5 atm for 5 min [45], while pressurized hot water extraction (PHWE) yielded 0.503 and 0.204 mg curcuminoids/g *C. longa* at 90 and 250°C, respectively [35] although *C. longa* contains higher curcuminoids, i.e. ca. 4.4% of dry weight, than *C. zedoaria*.

The neutral, ionic and basic types of NADESs give more or less similar yields of curcuminoids, while the lowest yields were obtained by acidic types of NADESs, i.e. CAS-H_2O (1:2:15) and MAS-H_2O (1:1:11) with yields of 0.151 ± 0.001 and 0.131 ± 0.002 mg/g, respectively. Though CCMA-H_2O (1:1:2) yielded the highest curcuminoids, apparently there is no direct relation between solubilizing capacity of the NADES with respect to curcumin and the

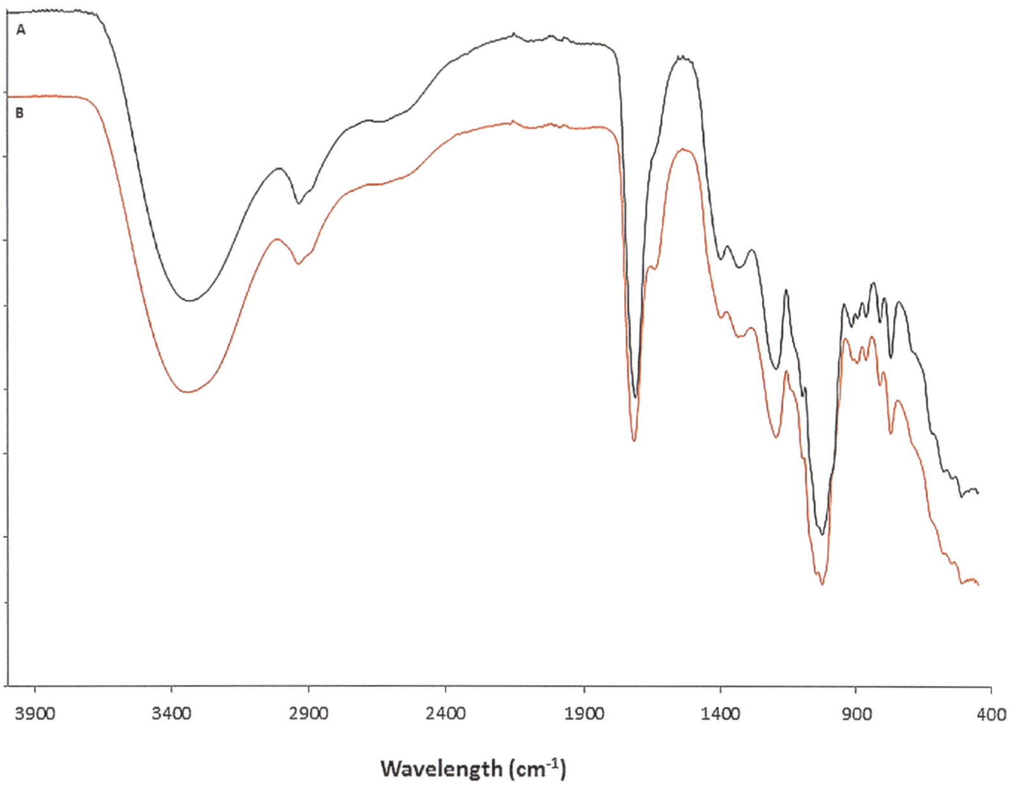

Figure 8. FT-IR spectra: (A) clear and transparent CAS-H_2O and (B) yellowish CAS-H_2O.

polarity, water content or pH [40]. However, exhaustive extraction of *C. zedoaria* using Soxhlet and maceration with ethanol (96%) as solvent only gave 0.119 ± 0.0001 and 0.152 ± 0.010 mg curcuminoids/g dry weight, respectively. Degradation of curcumin might occur at high temperature (78°C) during Soxhlet extraction. Salem et al. [46] reported that curcumin degraded after 24 h exposure at 70°C.

Longer curcuminoid extraction with CCMA-H_2O (1:1:2) up to 96 h (4 days) only gave a yield of 0.233 mg/g ± 0.017, i.e. approximately 36% less than that obtained after 24 h extraction. Though curcumin is reported to be stable at 10–55°C [46], prolonged exposure at 40°C (ca. 96 h) could degrade curcumin. In addition, CCMA-H_2O (1:1:2) could not stabilize curcumin, and extraction time may affect the obtained yield. In fact curcumin is precipitated after about 3 days due to its low water solubility at pH = 1–7 [47]. The CCMA-H_2O (1:1:2) had a pH of 2 at 10x dilution with Aquadest. The native pH of CCMA-H_2O (1:1:2) cannot be measured due to its inherent high viscosity.

In NADES with higher water content such as FS-H_2O (2:1:26) (water content = 40% by weight), only 37–52% curcuminoids were left after 96 h. This coincides with Tonnesen and Karlsen [47] who reported that the degradation of curcumin will be faster approximately 100x than that in concentrated solution at pH < 7. It clearly concludes that water content may affect the stabilizing

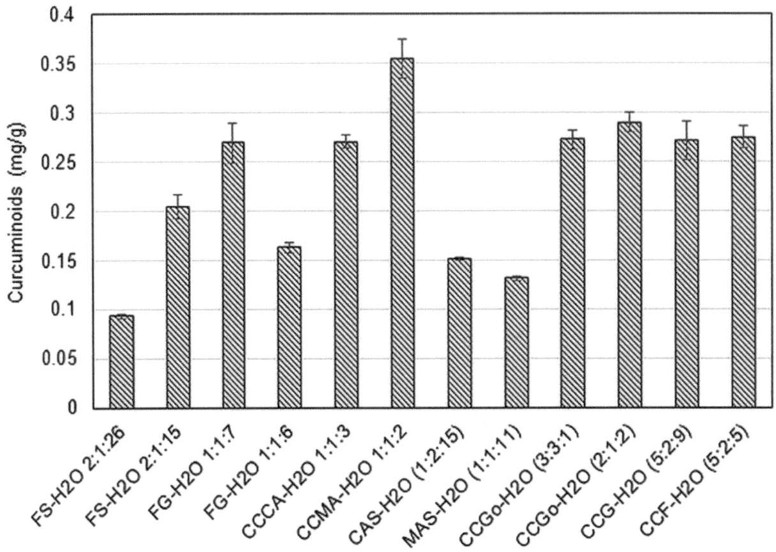

Figure 9. Curcuminoids extracted (mg curcuminoids/g dry weight) from *Curcuma zedoaria* with different types of NADES (G = glucose, F = fructose, S = sucrose, CC = choline chloride, MA = malic acid, CA = citric acid and Go = glycerol).

ability of NADES to curcuminoids. Moreover, the low level of curcuminoids might also be due to the loss of NADES structure since the hydrogen bond will rupture when NADES is further diluted with water content higher than 50% [20]. Though higher water content minimized the mass transfer resistance between NADES and plant matrix, water dilution affects considerably the target compounds and the nature of NADES itself.

A profile of extracted curcuminoids was also studied to find the selectivity of NADES to curcuminoids (**Table 4**). It is shown that NADES and FS-H_2O (2:1:26) are less selective to bis-demethoxyxurxumin (BDMC), but it is more selective to curcumin (ca. 54 wt.%) than to DMC (ca. 32 wt.%). Other kinds of NADESs also give different selectivity. The exhaustive extraction with 96% ethanol gave a similar selectivity, i.e. approximately 17, 23 and 60 wt.% for BDMC,

Extraction method	Yield (mg/g [a]) (% [b])			Extracted curcuminoids (mg/g ± SD)
	BDMC[c]	DMC	C	
Soxhlet (EtOH 96%)	0.020 (17)	0.028 (23)	0.071 (60)	0.120 ± 0.020
Maceration (EtOH 96%)	0.026 (18)	0.034 (21)	0.092 (61)	0.150 ± 0.660
FS-H_2O (2:1:26)	0.013 (14)	0.030 (32)	0.051 (54)	0.093 ± 0.002
CCMA-H_2O (1:1:2)	0.017 (5)	0.072 (20)	0.266 (75)	0.355 ± 0.019

[a]Yield expressed with mg bioactive compound/g dry weight of *C. zedoaria*.
[b]Percentage of weight to the total weight of extracted *curcuminoids* (*curcuminoid* = BDMC + DMC + C).
[c]*Bisdemethoxycurcumin* (BDMC), *demethoxycurcumin* (DMC) and *curcumin* (C).

Table 4. Profile of the extracted ***curcuminoids*** (mg/g) with different extraction methods.

http://dx.doi.org/10.5772/intechopen.71755

Trial	Pressurized extraction conditions						NADES[c]	Results
	Plant material[a]	Weight (mg)	Preheat (min)	Holding (min)[b]	Temp (°C)	Press (bar)		
E1 [d]	Pwd	100	30	15	50	100	CHCA (1:1) GMA (1:1) LPrS (2:1)	Clogging
E2 [d]	Frz	100	30	15	50	100		
E3 [d]	Pwd	300	60	2	50	50	MAS (1:1) CAS (1:1) GMA (1:1)	Clogging
E4 [d]	Pwd	200	60	2	50	50	CHCA (1:1) βAMA (1:1) LPrS (2:1)	Clogging
E5 [e]	Pwd	200	60	2	50	50	CHCA (1:1) GMA (1:1) LPrS (2:1)	Clogging
E6 [f]	Pwd	200	60	2	50	50	MAS (1:1) CAS (1:1) βAMA (1:1)	None

[a]Plant material Pwd = powder, Frz = freeze-dried. All the materials have 25–53 μm of particle size. Plant material and NADES were mixed prior loaded to the extractor cells.

[b]Heat up and discharge time is in default setting. They were 1 and 5 min, respectively. All the experiments were conducted with two cycles.

[c]Ratio is in molar.

[d]Plant material, NADES and 1 g of sea sand were mixed prior loaded to the extractor cells. There were flushing with solvent (water) and gas (N_2) between cycles, 1 min each. Around 3 g of sea sand was placed in the upper and lower part of the extractor cells.

[e]Plant material, NADES and 1 g of sea sand were mixed prior loaded to the extractor cells. No flush with solvent (only with gas 1 min) between cycles. Around 3 g of sea sand was placed in the upper and lower part of the extractor cells.

[f]Plant material, NADES and 1,5 g of sea sand were mixed prior loaded to the extractor cells. No flush with solvent (only with gas 1 min) between cycles. Around 2.5 g of sea sand was placed in the upper and lower part of the extractor cells.

Abbreviations: β-alanine (**βA**), citric acid (**CA**), choline chloride (**CH**), glucose (**G**), malic acid (**MA**), L-Proline (**LPr**), sucrose (**S**).

Table 5. Preliminary extraction of pressurized extraction for finding the best NADES extraction conditions.

DMC and C, respectively. Hence, curcuminoid selectivity depends on the types of solvent. Moreover, in the case of curcuminoid extraction from *C. zedoaria* powder, the ratio of solid to NADES also affected the yield (data not shown). In conclusion, NADES is a better solvent to solubilize curcuminoids than water and ethanol. NADES is more selective to curcumin followed by DMC, while ethanol only gave 60% selectivity to curcumin.

The extraction of an alkaloid, galantamine, with NADES from its plant matrix *Narcissus pseudonarcissus sp. was* also conducted. A preliminary and simple extraction method similar to that of *C. zedoaria* explained above was also conducted with ground powder of *N. pseudonarcissus* with particle size around 25–53 μm. However, none of the galantamine was extracted although the extraction time was prolonged up to a week. This is probably because the mechanical stirring could not overcome the mass transfer resistance around the pellet of *N. pseudonarcissus* due to the structure of plant matrix.

Therefore, a pressurized extraction method using a pressurized extractor apparatus E-916 (Büchi, Flawil, Switzerland) was performed. It is a fast, simple and reproducible method facilitated by high-pressure condition instead of mechanical agitation. In addition, it is also a fast screening method to find the best extraction condition for NADES extraction. Preliminary experiments were conducted to find the best extraction configuration, whilst NADESs are used as a solvent as shown in **Table 5**.

Two different kinds of plant matrix were used, i.e. ground powder and freeze-dried powder of *N. pseudonarcissus* bulb. Both have particle sizes of 25–53 μm. NADES with mild viscosity was chosen to minimize clogging problem inside the extraction cell which is essentially a packed bed where the bulk porosity is important to make a good contact between NADES and plant matrix. If the sample inside the extraction cell is too compact, it creates a clogging problem and overpressurized the cell; otherwise the contact between the solute and solvent will be minimal leading to a low yield.

At high-pressure conditions (100 bar), E1 and E2, NADESs were overcooked, and the sugar was caramelized although only 30 min of preheating time was applied. Therefore, the extractions were performed at a lower pressure (50 bar). To compensate the pressure reduction, the preheating time was increased. Finally, extraction condition E6 was found to be free of clogging and overcooked problems. Further extraction of galantamine with NADES by applying E6 condition gave 6.11 and 9.33 mg of galantamine/g dry weight for CAS (1:1) and MAS (1:1), respectively. These yields were higher than that obtained by water extraction as a control (5.35 mg of galantamine/g dry weight). These values were also higher than galantamine extraction with supercritical CO_2, i.e. 303 μg/g dry weight (70°C, 220 bar, 3 h) [48]. The selectivity of NADES to galantamine was slightly better (70–78%) than that of SC-CO_2 (<70%). Thus, NADES extraction of galantamine is more efficient, in terms of both galantamine yield and selectivity.

5. Concluding remarks

Deep eutectic solvents (DESs), including natural deep eutectic solvents (NADESs), are new generation of solvents. DES can be prepared by mixing two or more components at eutectic composition so that the mixture has lower melting point than those of the constituent components. DESs are typically prepared from quaternary ammonium halide salts and hydrogen bond donors (HBDs), such as amide, carboxylic acid, alcohol, sugar, amino or organic acid and alkylsulfate or alkyl phosphate. The behaviors and properties of DESs can be adjusted by varying the hydrogen bond donors and their molar ratio in the mixtures [13]. The polarities of DESs can match those of conventional organic solvents; however, DESs have several advantages, such as non-toxic, non-volatile, inflammable and biodegradable. Therefore, DESs can be considered as green solvents.

DESs potentially have unlimited number of applications. The applications of DESs to purify crude biodiesel made from rice bran oil and to extract natural bioactive compounds, such as oryzanol, cucurmin and galantamine, have been discussed in this chapter. DESs could give higher extraction yields of natural bioactive compounds than those obtained using conventional organic solvents. Besides that, the use of DESs could lead to a simpler extraction or separation procedure, especially in a large industrial scale, due to non-toxic and inflammable nature of DES.

However, at the same time, one has to consider that there is no universal solvent for all kinds of compounds. Each material and targeted compound requires the development of a specific process. No single standard procedure of extraction is suitable for extractions of all secondary metabolites. This also allows a rather selective extraction which is particularly of interest for the isolation of pure compounds. Some parameters that should be considered in the development of a DES-extraction procedure are types of the matrix plant, type of extraction process, ratio of plant material to solvent and extraction conditions: temperature, duration of extraction, viscosity and water content of DES.

Author details

Siti Zullaikah[1]*, Orchidea Rachmaniah[1], Adi Tjipto Utomo[1], Helda Niawanti[2] and Yi Hsu Ju[3]

*Address all correspondence to: szulle@chem-eng.its.ac.id

1 Department of Chemical Engineering, Institut Teknologi Sepuluh Nopember, Kampus ITS Keputih Sukolilo, Surabaya, Indonesia

2 Department of Chemical Engineering, Mulawarman University, Samarinda, Indonesia

3 Department of Chemical Engineering, National Taiwan University of Science and Technology, Taipei, Taiwan

References

[1] Abidin MHZ, Hayyan M, Hayyan A, Jayakumar NS. New horizons in the extraction of bioactive compounds using deep eutectic solvents: A review. Analytica Chimica Acta. 2017;**979**:1-23

[2] Vieira FA, Guilherme RJR, Neves MC, Abreu H, Rodrigues ERO, Maraschin M, Coutinho JAP, Ventura SPM. Single-step extraction of carotenoids from brown macroalgae using nonionic surfactants. Separation and Purification Technology. 2017;**172**:268-276

[3] Wang L, Weller CL. Recent advances in extraction of nutraceuticals from plants. Trends in Food Science and Technology. 2006;**17**:300-312

[4] Deetlefs M, Seddon KR. Assessing the greenness of some typical laboratory ionic liquid preparations. Green Chemistry. 2010;**12**:17-30

[5] Zhang Q, Vigier KDO, Royer S, Jerome F. Deep eutectic solvents: Syntheses, properties and applications. Chemical Society Reviews. 2012;**41**:7108-7146

[6] Zhang S, Ye L, Zhang H, Hou J. Green-solvent-processable organic solar cells. Materials Today. 2016;**19**:533-543

[7] Almeida MR, Passos H, Pereira MM, Lima AS, Coutinho JAP, Freire MG. Ionic liquids as additives to enhance the extraction of antioxidants in aqueous two-phase systems. Separation and Purification Technology. 2014;**128**:1-10

[8] Gorke J, Srienc F, Kazlauskas R. Toward advanced ionic liquids. Polar, enzyme-friendly solvents for biocatalysis. Biotechnology and Bioprocess Engineering. 2010;**15**:40-53

[9] Wood N, Stephens G. Accelerating the discovery of biocompatible ionic liquids. Physical Chemistry Chemical Physics. 2010;**12**:1670-1674

[10] Domínguez de María P, Maugeri Z. Ionic liquids in biotransformations: From proof of-concept to emerging deep-eutectic-solvents. Current Opinion in Chemical Biology. 2011;**15**: 220-225

[11] Pham TPT, Cho CW, Yun YS. Environmental fate and toxicity of ionic liquids: A review. Water Research. 2010;**44**:352-372

[12] Kareem MA, Mjalli FS, Hashim MA, Alnashef IM. Phosphonium-based ionic liquids analogues and their physical properties. Journal of Chemical & Engineering Data. 2010;**55**:4632-4637

[13] Abbott AP, Boothby D, Capper G, Davies DL, Rasheed RK. Deep eutectic solvents formed between choline chloride and carboxylic acids: Versatile alternatives to ionic liquids. Journal of the American Chemical Society. 2004;**126**:9142-9147

[14] Abbott AP, Capper G, Davies DL, Rasheed RK, Tambyrajah V. Novel solvent properties of choline chloride/urea mixtures. Chemical Communications. 2003;**1**:70-71

[15] Smith EL, Abbot AP, Ryder KS. Deep eutectic solvent (DESs) and their applications. Chemical Reviews. 2014;**114**:11060-11082

[16] Gutiérrez MC, Ferrer ML, Mateo CR, del Monte F, Freeze-drying of aqueous solutions of deep eutectic solvents: A suitable approach to deep eutectic suspensions of self-assembled structures, Langmuir. 2009;**25**:5509-5515

[17] Imperato G, Eibler E, Niedermaier J, Konig B. Low-melting sugar-urea-salt mixtures as solvents for Diels-Alder reactions. Chemical Communications. 2005;**9**:1170-1172

[18] Francisco M, van den Bruinhorst A, Kroon MC. Low-transition-temperature mixtures (LTTMs): A new generation of designer solvents, Angewandte Chemie International Edition. 2013;**52**:3074-3085

[19] Choi YH , van Spronsen J, Dai y, Verberne M, Hollmann F, Arends IWCE , Witkamp GJ, verpoorte R. Verpoorte, Are natural deep eutectic solvents the missing link in understanding cellular metabolism and physiology? Plant Physiology. 2011;**156**:1701-1705

[20] Dai Y, van Spronsen J, Witkamp GJ , Verpoorte R . Choi YH. Natural deep eutectic solvents as a new potential media for green technology. Analytica Chimica Acta. 2013;**766**:61-68

[21] Ju YH, Vali SR. Rice bran oil as a potential resource for biodiesel: A review. Journal of Scientific and Industrial Research. 2005;**64**:866-882

[22] Ju YH, Zullaikah S. Effect of acid-catalyzed methanolysis on the bioactive components of rice bran oil. Journal of the Taiwan Institute of Chemical Engineers. 2013;**44**:924-928

[23] Zullaikah S, Lai CC, Vali SR, Ju YH. A two-step acid-catalyzed process for the production of biodiesel from rice bran oil. Bioresource Technology. 2005;**96**:1889-1896

[24] Lin L, Ying D, Chaitep S, Vittayapadung S. Biodiesel production from crude rice bran oil and properties as fuel. Applied Energy. 2009;**86**:681-688

[25] Shiu PJ, Gunawan S, Hsieh WH, Kasim NS, Ju YH. Biodiesel production from rice bran by a two-step in-situ process. Bioresource Technology. 2010;**101**:984-989

[26] Zullaikah S, Rahkadima YT, Ju YH. A non-catalytic in situ process to produce biodiesel from a rice milling by-product using a subcritical water-methanol mixture. Renewable Energy. 2017;**111**:764-770

[27] European Committee for Standardization (CEN). EN14214:2003. Automotive fuels–fatty acid methyl esters (FAME) for diesel engines—requirements and test methods. 2003

[28] Perkins SL, Painter P, Colina CM. Experimental and computational studies of choline chloride-based deep eutectic solvents. Journal of Chemical and Engineering Data. 2014;**59**:3652-3662

[29] Wagle DV, Deakyne CA, Baker GA. Quantum chemical insight into the interactions and thermodynamics present in choline chloride based deep eutectic solvent. The Journal of Physical Chemistry B. 2016;**120**:6739-6746

[30] Shahbaz K, Mjalli FS, Hashim MA, AlNashef IM. Using deep eutectic solvents for the removal of glycerol from palm oil-based biodiesel. Journal of Applied Sciences. 2010;**10**:3349-3354

[31] Hayyan M, Aissaoui T, Hashim MA, AlSaadi MAH, Hayyan A. Triethylene glycol based deep eutectic solvents and their physical properties. Journal of the Taiwan Institute of Chemical Engineers. 2015;**50**:24-30

[32] Aissaoui T. Novel contribution to the chemical structure of choline chloride based deep eutectic solvent. Pharmaceutica Analytica Acta. 2015;**6**:2153-2435

[33] Smith BC. Infrared spectral interpretation: A systematic approach. CRC Press United States. 1998

[34] Niawanti H, Zullaikah S. Removal of bioactive compound (γ-Oryzanol) from Rice bran oil-based biodiesel using deep eutectic solvent. Chemical Engineering Transaction. 2017;**56**:1513-1518

[35] Euterpio MA, Cavaliere C, Capriotti AL, Carlo C. Extending the applicability of pressurized hot water extraction to compounds exhibiting limited water solubility by pH control: Curcumin from the turmeric rhizome. Analytical and Bioanalytical Chemistry. 2011;**401**:2977-2985

[36] Dai Y, Verpoorte R., Choi YH. Natural deep eutectic solvents providing enhanced stability of natural colorants from safflower (Carthamus tinctorius). Food Chemistry. 2014;**159**:116-121

[37] Bajkacz S, Adamek J. Evaluation of new natural deep eutectic solvents for the extraction of isoflavones from soy products. Talanta. 2017;**168**:329-335

[38] Xin R, Qi S, Zeng C, Khan FI, Yang B, Wang Y. A functional natural deep eutectic solvent based on trehalose: Structural and physicochemical properties. Food Chemistry. 2017;**217**:560-567

[39] Huang D, Zhang S, Zhong T, Ren W, Yao X, Guo Y, Duan XC, Yin YF, Zhang SS, Zhang X. Multi-targeting NGR-modified liposomes recognizing glioma tumor cells and vasculogenic mimicry for improving anti-glioma therapy. Oncotarget. 2016;**7**:43616-43628

[40] Wikene KO, Bruzell E, Tønnesen HH. Characterization and antimicrobial phototoxicity of curcumin dissolved in natural deep eutectic solvents. European Journal of Pharmaceutical Sciences. 2015;**80**:26-32

[41] Radošević K, Ćurko N, Srček VG, Bubalo MC, Tomašević M, Ganić KK, Redovniković IR. Natural deep eutectic solvents as beneficial extractants for enhancement of plant extracts bioactivity. LWT-Food Science and Technology. 2016;**73**:45-51

[42] Clarke MA, Edye LA, Eggleston G. Sucrose decomposition in aqueous solution, and losses in sugar manufacture and refining. Advances in Carbohydrate Chemistry and Biochemistry. 1997;**52**:441-470

[43] Quintas M, Guimarães C, Baylina J, Brandão TRS, Silva CLM. Multiresponse modelling of the caramelisation reaction. Innovative Food Science & Emerging Technologies. 2007;**8**:306-315

[44] Patra D, Sleem F. A new method for pH triggered curcumin release by applying poly (L-lysine) mediated nanoparticle-congregation. Analytica Chimica Acta. 2013;**795**:60-68

[45] Kwon HL, Chung MS. Pilot-scale subcritical solvent extraction of curcuminoids from curcuma long L. Food Chemistry. 2015;**185**:58-64

[46] Salem M, Rohani S, Gillies ER. Curcumin, a promising anti-cancer therapeutic: A review of its chemical properties, bioactivity and approaches to cancer cell delivery. Royal Society of Chemistry Advance. 2014;**4**:10815-10829

[47] Tønnesen HH, Karlsen J. Studies on curcumin and curcuminoids, Zeitschrift für Lebensmittel-Untersuchung und. Forschung. 1985;**180**:402-404

[48] Rachmaniah O, Choi YH, Arruabarrena I, Vermeulen B, Verpoorte R, van Spronsen J, Verpoorte R, Witkamp GJ. Environmentally benign super critical CO2 extraction of galantamine from floricultural crop waste of Narcissus Pseudonarcissus. The Journal of Supercritical Fluids. 2014;**93**:7-19

Chapter 3

Solventless Extraction of Essential Oil

Muhammad Shahzad Aslam,
Muhammad Syarhabil Ahmad and Sarwat Ali Raja

Additional information is available at the end of the chapter

http://dx.doi.org/10.5772/intechopen.72401

Abstract

Essential oil is one of an important concentrated liquid that possesses many physical, chemical and pharmacological properties. Extraction of essential is one of the main issues in the last decade. Conventional treatment consisting of hydrodistillation and steam distillation has many disadvantages and finds difficult to purify essential oil. Now, it is much easier to extract essential oil with the invention of new greener technologies that reduce the involvement of solvent, decrease the extraction time, energy and descent the interaction of the concentrated volatile liquid with atmospheric oxygen through the application of vacuum.

Keywords: solventless extraction, solvent-free extraction, essential oil, extraction technology, green extraction

1. Introduction

Essential oils are complex concentrated liquids comprise of volatile compounds. They have been extracted from numerous plant [1]. They have been widely used as a food preservative (eucalyptus essential oil, thyme), cosmetic preparation (lavender oil), antimicrobial (lemon grass, cumin, fennel), and anticancer agent (lemon grass, *Croton flavens*). Hydrodistillation and steam distillation is the common conventional method of extracting the essential oil [2–9]. These methods have some disadvantages Preservation of essential oil from its environment can be possible through a number of technologies such as nanospheres, liposome, microcapsules and nanoemulsions [10]. A lot of research is currently underway for extracting essential oil through new green methods. This review chapter documents the updated information about novel solvent-less extraction of essential oil.

2. Solventless extraction in a closed system

Solventless extraction of essential oil was designed in a closed system under reduced pressure using a vacuum and compared the results with the conventional methods such as hydrodistillation. The quality and quantity of essential oil extracted from *Chromalaena odorata, Citronella, Baeckea frutescens,* Orange peel was found better than traditional method [11, 12]. The novel closed system was also optimized using central composite design (CCD) on *Chromalaena odorata* and maximum extraction yield was found to be under an ideal condition at 80°C temperature and 8 h of time [13].

3. Solvent-free microwave extraction

Solvent-free microwave extraction (SFME) of volatile natural substances was the first to patent in 2004. Farid Chemat et al. invented a method of extraction of essential oil consisting of a microwave oven with a microwave chamber for receiving the biological material and a condensation chamber. It was first tested on different spices such as ajowan (*Carum ajowan,* Apiaceae), cumin (*Cuminum cyminum,* Umbelliferae), star anise (*Illicium anisatum,* Illiciaceae) and the result had published on 4 February 2004. This technique is more fast, without solvent and effective when compared to hydrodistillation [14]. It was also evaluated on three aromatic herbs basil (*Ocimum basilicum*), garden mint (*Mentha crispa*), and thyme (*Thymus vulgaris*) where extraction time was significantly decreased from 4.5 h (hydro-distillation) to 30 min (SFME) [15]. This method was modified by many scientists. SFME method was modified by Wang et al. on dried *Cuminum cyminum* L. and *Zanthoxylum bungeanum* Maxim. by adding carbonyl iron powders (CIP) and mixed with the sample. CIP helps to reduce time (30 min) and microwave power (85 W) [16]. An attempt had been made to improve solvent-free extraction of essential oil using graphite powder, activated carbon powder. The effect was studied on *Illicium verum* Hook. f. and *Zingiber officinale* Rosc. [17]. Pressurized solvent-free microwave assisted extraction was used for extraction of phenolic compounds. The best extraction conditions were obtained, in a laboratory scale extractor of 50 mL filled with 4 g fresh berries, using a 1000 W microwave power applied during 50 s and repeated 5 cycles [18]. Solvent-free microwave extraction was modified by changing the flow of product toward the gravitation force. It is also known as Microwave dry-diffusion and gravity. It was better than hydrodistillation. The extraction performed in just 45 minutes with less energy and a clean process [19]. The extraction condition of SFME was optimized on *Elettaria cardamomum* (L.) using central composite design (CCD). The conditions such as time (min), power (W), humidity (%) was optimized by (CCD) and percentage yield (%) of was compared [20]. Optimization of SFME of pigeon pea leaves performed on an aliquot of 200 g plant materials that were wetted before extraction by soaking in water for 1 h [21]. Optimum parameters of SFME was performed on *S. chinensis* fruits and found ideal extraction time of 30 min, irradiation power 385 W and the moisture content of 68% respectively [22]. The quality of essential oil was also evaluated using Solvent-free microwave extraction method and compared with conventional method. It was found to be more effective than conventional method [23]. The effect of solvent-free microwave extraction was performed on several medicinal plants such as *Calamintha nepeta* [24], Basil leaves [25], *Dryopteris fragrans* [26], *Schisandra chinensis* [27], *Cymbopogon winterianus* [28].

Medicinal plants	Identified essential oil	Pharmacological activities	Reference
Cuminum cyminum L. and *Zanthoxylum bungeanum* Maxim	2-Methyl-5-(1-methylethyl)-bicyclo[3.1.0]hex-2-ene, 1-Methylethylideneyl-cyclohexane, α-Pinene, Camphene, β-Phellandrene, β-Pinene, β-Myrcene, β-Phellandrene 3-Carene, 4-Carene, 1-Methyl-2-(1-methylethyl) benzene, d-Limonene, Eucalyptol, 1-Methyl-4-(1-methylethyl)-1,4-cyclohexadiene, 1-Methyl-4-(1-methylethylidene)-cyclohexene, β-Terpineol, 6,6-Dimethyl-2-methylene-bicyclo[2.2.1]heptan-3-one, 4-Methyl-1-(1-methylethyl)-3-cyclohexen-1-ol, Pulegone, Cuminal 4-(1-Methylethyl)-1-cyclohexene-1-carboxaldehyde 2-Ethylidene-6-methyl-3,5-heptadienal, α-Proyl-benzenemethanol, 4-(1-Methylethyl)-1,4-cyclohexadiene-1-methanol, 6-Isopropylidene-1-methyl-bicyclo[3.1.0]hexane, Caryophyllene, 2,6-Dimethyl-6-(4-methyl-3-pentenyl)-bicyclo[3.1.1]hept-2-ene, 7,11-Dimethyl-3-methylene-1,6,10-dodecatriene, 2-Isopropyl-5-methyl-9-methylene-bicyclo[4.4.0]dec-1-ene, Octahy-dro-3,8,8-trimethyl-6-methylene-1H-3a,7-methanoazulene, Thujopsene, 1-(1,5-Dimethyl-4-hexenyl)-4-methylbenzene 5-(1,5-Dimethy-4-hexenyl)-2-methyl-1,3-cyclohexadiene, Copaene, 1-Methyl-4-(5-methyl-1-methylene-4-hexenyl-cyclohexene, β-Sesquiphellandrene, Caryophyllene oxide, Carotol	N.F	[16]
Rosmarinus officinalis L.	a-Pinene, Camphene, b-Pinene, Myrcene, a-Phellanderene, a-Terpinene c-Terpinene, Linalool, 1 1,8-Cineole, Camphor, Borneol, b-Caryophyllene, Trans b-ocimene, cis-Sabinene hydrate, Verbinone, Terpene-4-ol, Myrtenol, Bornyl acetate, Cis-jasmone, a-Humulene, Pentasiloxane Caryophyllene, 1,5-Diphenyl 2H-1,2,4 triazoline, 1-Methyl-2,4-nitrophenyl benzimid, 2-Methoxy-3,8-dioxocephalotax-1-ene, 1,2-Benzenedicarboxylic acid, 9-Octadecenoic acid, Docosanoic acid	Anti-bacterial	[31]
Dryopteris fragrans	Cedrene [1S-(1a,4a,7a)]-1,2,3,4,5,6,7,8-octahydro-1,4,9,9-tetramethyl-4,7-methanoazulene Caryophyllene, 10S,11S-himachala-3(12),4-diene, 4-(2,6,6-Trimethyl-2-cyclohexen-1-yl)-2-butanone, (E)-4-(2,6,6-trimethyl-2-cyclohexen-1-yl)-3-buten-2-one, 1,2,3,4,4a,5,6,8a-Octahydro-7-methyl-4-methylene-1-(1-methylethyl)naphthalene, (R)-c-cadinene 4-(2,6,6-Trimethyl-1-cyclohexen-1-yl)-3-buten-2-one, (1a,4ab,8aa)-1,2,3,4,4a,5,6,8a-Octahydro-7-methyl-4-methylene-1-(1-methylethyl)naphthalene, Albicanol, [1R-(1a,4ab,8aa)]-decahydro-1,4a-dimethyl-7-(1-methylethylidene)-1-naphthalenol, Calarene epoxide, (1R,4S,11R)-4,6,6,11-tetramethyltricyclo[5.4.0.0(4,8)] undecan-1-ol, 1,4,4a,5,6,7,8,8a-Octahydro-2,5,5,8a-tetramethyl-1-naphthalenemethanol, (-)-Isolongifolol, acetate, 1R,4S,7S,11R-2,2,4,8 tetramethyltricyclo[5.3.1.0(4,11)]undec-8-ene, [3S-(3a,5a,8a)]-1,2,3,4,5,6,7,8-octahydro-a,a,3,8-tetramethyl-5-azulenemethanol acetate,	Anti-oxidant	[26]
Schisandra chinensis (Turcz.) Baill	a-Pinene, Camphene, 2-Carene, D-Limonene, o-Cymene, gamma-Terpinene, Thymol methyl ether, L-Bornyl acetate, Cyclocopacamphene, Ylangene, beta-bourbonene, (+)-Sativene,(−)-beta-Elemene, Germacrene-D, (E)-(b)-Farnesene, (E)-a-bergamotene, Elixene, alpha-amorphene,(−)-beta-Chamigrene, Beta-bisabolene, b-maaliene, c-cadinene, b-himachalene, a-Chamigrene, (+)-a-Longipinene, (+)-Cuparene, b-Caryophyllene, Guaiene, (+)-Ledene, L-calamenene	Anti-oxidant	[22]
Cajanus cajan (L.) Millsp.	3,6-Dimethyl-octane, Naphthalene, Dodecane, 6-Ethyl-undecane, 4-Methyl-dodecane, 4-Ethyl-undecane, 4,6-Dimethyl-dodecane, 1-Methyl-naphthalene, 2,6,11-Trimethyl-dodecane, α-Longipinene, 2-Methyl-tridecane, (+)-Cyclosativene, Ylangene, α-Copaene, Tetradecane, Longifolene, Caryophyllene, β-Selinene, α-Bergamotene, α-Himachalene, Humulene, Alloaromadendrene, α-Bisabolene, 2,4-Bis(1,1-dimethylethyl)-phenol, δ-Cadinene, Hexadecane, Norphytane	Anti-microbial activities	[21]

Medicinal plants	Identified essential oil	Pharmacological activities	Reference
Ocimum basilicum L.	Sabinene, Octen 3 ol, β-Pinene, Heptanol, β-Myrcene, p-Cymene, Limonene, 1,8-Cineole, β-Ocimene, γ-Terpinene, Fenchone, Linalool, Camphor, Menthol, α-Terpineol, Methyl chavicol, Nerol, Neral, Geraniol, Geranial, α-Terpinenyl acetate, Neryl acetate, α-Copaene, Geranyl acetate, β-Bourbonene, β-Cubebene, β-Elemene, Methyl eugenol, β-Caryophyllene, α-Bergamotene, α-Humulene, β-Farnesene, Germacrene-D, γ-Cadinene, Δ-Cadinene, α-Bisabolene, β-Bisabolene, Spathulenol, Caryophyllene oxide, α-Cadinol	N.A	[25]
Hippophae rhamnoides	Isorhamnetin, isorhamnetin 3-O-glucoside, isorhamnetin 3-O-rutinoside and quercetin 3-O-glucoside	Anti-oxidant	[32]
Lavandula angustifolia	1, 8-cineole, Camphor, Borneol, p-cymene, Limonene, Cryptone, isobornyl formate, cumin aldehyde, Valerianol, α-pinene	Anti-bacterial activity	[33]
Calamintha nepeta	a-Thujene, a-Pinene, Sabinene, b-Pinene, D2-Carene, a-Terpinene, Camphene, (Z)-b-Ocimene, allo-Ocimene, Myrcene, Limonene, c-Terpinene, p-Cymene, Octan-3-ol, Eugenol, Geranyl acetone, Hexahydrofarnesylacetone, Octacosane, Phytol, Caryophyllene oxide, T-Cadinol, a-Cadinol, T-Muurolol, a-Copaene, b-Elemene, b-Cubebene, b-Bourbonene, a-Humulene, Caryophyllene, Germacrene-D, c-Cadinene, epi-Sesquiphellandrene, d-Cadinene, Menthyl acetate, Bornyl acetate, Menthol, Chrysanthenone, Piperitenone, Piperitenone oxide, Isopulegonen, Pulegone, Piperitone, cis-Sabinene hydrate, 1,8-Cineole, Dihydrocarveol, trans-Sabinene hydrate, Menthone, Isomenthone, Terpinen-4-ol, a-Terpineol	N.A	[24]

Table 1. List of medicinal plants and their identified essential oil evaluated under microwave assisted solvent-free extraction.

Instrumentation	Extraction conditions	Results	Reference
Reactor (500 mL), microwave oven, agitator, shielded non-invasive thermometry system, transformer with maximum output power is 800 W with 2450 MHz of microwave irradiation frequency (MIF).	100 g of sample and 20 g of carbonyl iron powder (CIP) were added inside the reactor, stirred, heated (85 W) for 30 min at 100°C with speed of rotation (200 rpm), concentrated outside the microwave oven by a cooling system	CIP helps to improve the microwave absorption capacity than water and is faster (30 min) than conventional method	[16]
The multi-mode reactor (2 × 800 W, 2450 MHz), rotating microwave diffuser, plasma coated PTFE cavity, circulating cooling system at 5°C	250 g of Rosmarinus leaves were placed into the reactor without the addition of water or any solvent	Higher amounts of oxygenated monoterpenes were found as compared to conventional method	[31]
Microwave-accelerated reaction system (1000 W, 2450 MHz) multimode microwave reactor armed with a TFT multicolour liquid crystal screen, a power sensor (power range 0–1000 W), an infrared temperature sensor, a temperature controller and electromagnetic stirrer	200 g plant material was moistened prior to extraction by soaking in certain proportions of water (weight basis) for 1 h and then draining off the excess water. After that, the moistened materials were subjected to the microwave oven cavity and a condenser was used to collect the extracted essential oils in a pre-setting procedure.	A maximal extraction yield of 0.33% was achieved under optimal conditions of extraction time 34 min, irradiation power 520 W and humidity 51% 16	[26]

Instrumentation	Extraction conditions	Results	Reference
Multimode microwave reactor, temperature Infrared sensor with maximum output power is 700 W, microwave rotating diffuser that ensures homogeneous microwave distribution	A 100 g of *S. chinensis* fruits were moistened prior extraction by soaking in water then draining the excess of water. The extraction was continued until no distillate was obtained. The essential oil was collected, dried over anhydrous sodium sulfate and stored at 0°C until analyzed	Identification of optimum parameters was as follows; extraction time 30 min, irradiation power 385 W and the moisture content of *S. chinensis* fruits 68%, respectively	[22]
The microwave-accelerated reaction system with multimode microwave reactor (2.45 GHz), IR temperature sensor, an electromagnetic stirrer, a time calculator controller, circulating water-cooling system	200 g plant materials were wetted before extraction by soaking in a certain proportion of water for 1 h, and then removal the excess water. The wetted material was placed in the reaction flask and connected to a glass reaction flask	The optimal parameters were extraction time 44 min, irradiation power 660 W, and humidity 68%, with extraction yield of 0.330 (%, w/w)	[21]
Microwave oven (EMM-2007X, Electrolux, 20 l, the maximum delivered the power of 800 W) with a wave frequency of 2450 MHz. A round bottom flask with a capacity of 1000 ml was placed inside the oven and was connected to the three-way adapter and Liebig condenser through the hole. Then, the hole was closed with PTFE to prevent any loss of the heat inside	150 g of fresh plant materials were placed in the reaction flask and heated by microwave irradiation with 400 W (50% power) for 30 min without adding any solvent or water. During the process, the vapor passed through the condenser outside the microwave cavity where it was condensed. Essential oil and water were simply separated by decantation. The essential oil was collected in amber vials, dried over anhydrous sodium sulfate and stored at 277 K	SFME exhibit shorter extraction times as compared to conventional method (30 min vs. 4.5 h) and better yields (0.13% vs. 0.11%)	[25]
Milestone EOS-G microwave laboratory oven having a multimode microwave reactor (2.45 GHz) with a maximum delivered the power of 900 W. The extraction vessels are made from Pyrex and have a capacity of 1000 mL with a temperature sensor optic fiber which was inserted in the center of embedded plant material and also in the reactor above the matrix	400 g of sea buckthorn press cake was heated using a fix power density 1 $W \cdot g^{-1}$ without the addition of solvents or water. The crude extract was collected continuously in a graduated cylinder. The extraction was continued until no more extract was obtained or overheating was detected	This method exhibit shorter extraction time (15 min), cleaner feature (no solvent or water used) and extraction of valuable flavonoids (Isorhamnetin, isorhamnetin 3-O-glucoside, isorhamnetin 3-O-rutinoside and quercetin 3-O-glucoside) at optimized power (400 W)	[32]
Microwave apparatus, 2450 MHz with maximum power 1000 W and ACTE0 sensor for temperature monitoring. The power of the oven was 500 W for 10 min. The temperature was achieved at 95°C, and the extraction was carried out for 25 min	30 g of dried *Lavandula angustifolia* was soaked in 20 mL distilled water at room temperature (25°C) for 1 h in order to hydrate the external layers of the plant material. The moistened plant material was placed in a flat-bottom flask combined with a Clevenger apparatus. The SFME process was performed for 35 min. The essential oils were collected in amber colored vials, dehydrated with anhydrous sodium sulfate, capped under nitrogen and kept at 4°C	It helps to extract more oxygenated compounds	[33]

Instrumentation	Extraction conditions	Results	Reference
Vacuum and nitrogen gas was applied on and off to remove air and replacing it with nitrogen in a closed system. At the end of the extraction process water and oil was separated and anhydrous sodium sulfate was used to dry the excess water	Fresh leaves of aromatic plants were grinded and to break them into smaller pieces and increasing the area of contact. Then, the grind leaves were put in a flask which was connected to another flask as a receiving flask. Firstly, the raw material was cooled down to a very low temperature to prevent decomposition and to avoid premature oil evaporation	Essential oil produced is lighter in color, higher yield, contains a cleaner, better purity and produced a stronger aroma compared to the essential oil produced from hydro-distillation	[12]

Table 2. Instrumentation and extraction conditions of solvent-free extraction.

This modern method was transformed from laboratory scale to pilot and industrial scale [29]. List of Medicinal plants, their identified essential oil evaluated under Microwave assisted solvent-free extraction were represented in **Table 1**. Instrumentation and extraction conditions of solvent-free extraction were mentioned in **Table 2** (**Figures 1–3**).

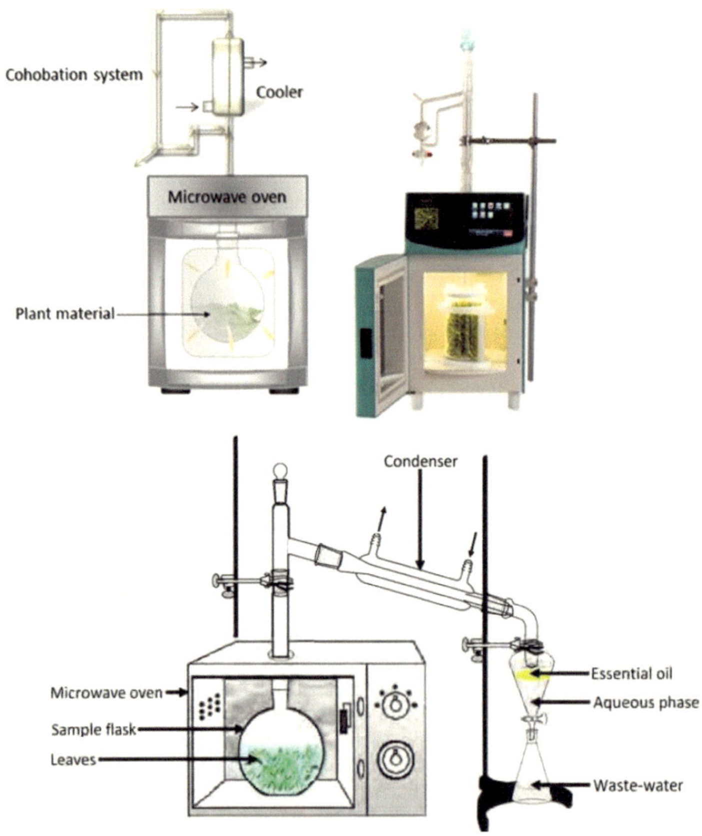

Figure 1. Solvent-free microwave extraction (SFME) [25, 30].

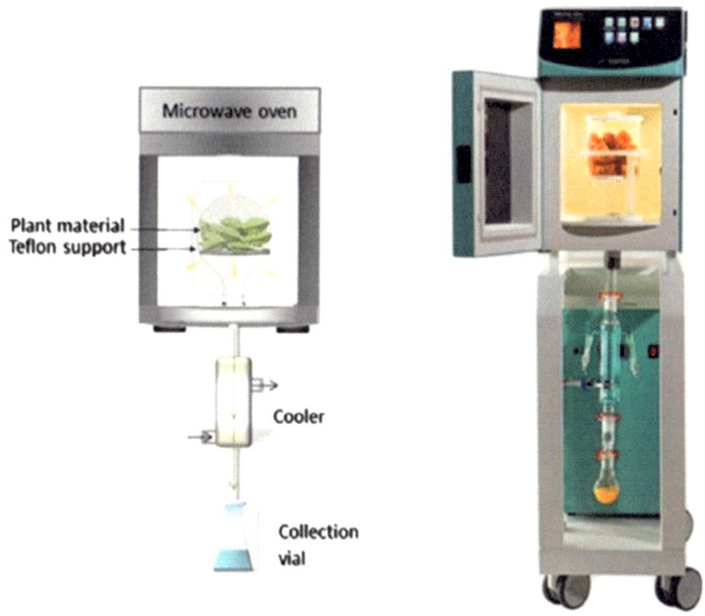

Figure 2. Microwave hydro-diffusion and gravity (MHG) [30].

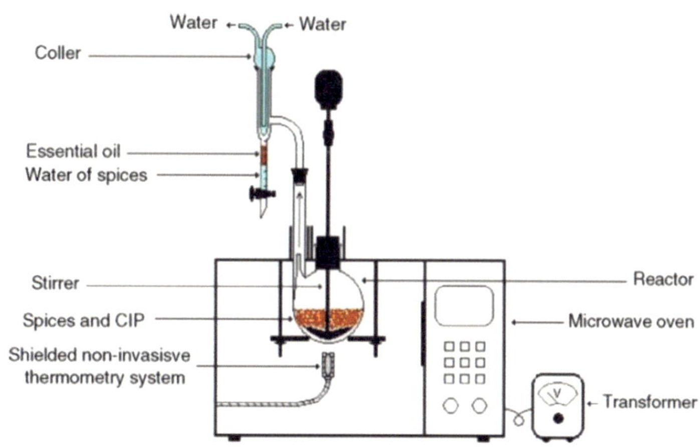

Figure 3. Improved SFME [16].

4. Conclusion

The term Solvent-less and solvent-free extraction have been used as synonymous with each other. Extraction of essential oil using these methods has a number of advantages such as fast action, cleanliness, green method, low energy output as compared to traditional extraction method. However, microwave extraction needs extra care before use as it may cause some negative effect on human health. There are many opportunities and modification possible in term of purification of essential oil by applying in combination with other extraction technique.

Author details

Muhammad Shahzad Aslam[1,2]*, Muhammad Syarhabil Ahmad[1] and Sarwat Ali Raja[3]

*Address all correspondence to: aslammuhammadshahzad@gmail.com; shahzad.aslam@rlmc.edu.pk

1 School of Bioprocess Engineering, Universiti Malaysia Perlis (UniMAP), Arau, Perlis, Malaysia

2 Rashid Latif College of Pharmacy, Lahore, Pakistan

3 Lahore Pharmacy College, (A project of LMDC), Lahore, Pakistan

References

[1] Tongnuanchan P, Benjakul S. Essential oils: Extraction, bioactivities, and their uses for food preservation. Journal of Food Science. 2014;**79**(7)

[2] Cook CM, Lanaras T. The Encyclopedia of Food and Health. Chapter: Essential Oils: Isolation, Production and Uses. In: Cabballero B, Finglas P, Toldra F, editors. Academic Press; pp. 552-557

[3] Cavanagh HMA, Wilkinson JM. Biological activities of lavender essential oil. Phytotherapy Research. 2002;**16**(4):301-308

[4] Sharma PR et al. Anticancer activity of an essential oil from *Cymbopogon flexuosus*. Chemico-Biological Interactions. 2009;**179**(2-3):160-168

[5] Kumar Tyagi A et al. Eucalyptus essential oil as a natural food preservative: *In Vivo* and *In Vitro* antiyeast potential. BioMed Research International. 2014;**2014**:1-9

[6] Gonçalves ND et al. Encapsulated thyme (*Thymus vulgaris*) essential oil used as a natural preservative in bakery product. Food Research International. 2017;**96**:154-160

[7] Tzortzakis NG, Economakis CD. Antifungal activity of lemongrass (*Cympopogon citratus* L.) essential oil against key postharvest pathogens. Innovative Food Science & Emerging Technologies. 2007;**8**(2):253-258

[8] Kedia A, Prakash B, Mishra PK, Dubey NK. Antifungal and antiaflatoxigenic properties of *Cuminum cyminum* (L.) seed essential oil and its efficacy as a preservative in stored commodities. International Journal of Food Microbiology. 2014;**168-169**:1-7

[9] Diao W-R, Hu Q-P, Zhang H, Xu J-G. Chemical composition, antibacterial activity and mechanism of action of essential oil from seeds of fennel (*Foeniculum vulgare* mill.). Food Control. 2014;**35**(1):109-116

[10] Majeed H et al. Essential oil encapsulations: Uses, procedures, and trends. RSC Advances. 2015;**5**(72):58449-58463

[11] Nasshorudin D, Ahmad MS, Mamat AS. Novel closed system extraction of essential oil: A green approach. American Journal of Biochemistry. 2016;**6**(6):145-148

[12] Ahmad MS, Nasshorudin D, Mamat AS. Novel closed system extraction of essential oil: Impact on yield and physical characterization muhammad. 2nd International Conference on Bioresource Technology. 2014;**75**:42-46

[13] Nasshorudin D, Ahmad MS, Mamat AS, Rosli S. Optimization study of *Chromalaena odorata* essential oil extracted using solventless extraction technique. AIP Conference Proceedings. 2015;**1660**

[14] Lucchesi ME, Chemat F, Smadja J. An original solvent free microwave extraction of essential oils from spices. Flavour and Fragrance Journal. Mar. 2004;**19**(2):134-138

[15] Lucchesi ME, Chemat F, Smadja J. Solvent-free microwave extraction of essential oil from aromatic herbs: Comparison with conventional hydro-distillation. Journal of Chromatography. A. Jul. 2004;**1043**(2):323-327

[16] Wang Z et al. Improved solvent-free microwave extraction of essential oil from dried *Cuminum cyminum* L. and *Zanthoxylum bungeanum* maxim. Journal of Chromatography. A. Jan. 2006;**1102**(1-2):11-17

[17] Wang Z et al. Rapid analysis of the essential oils from dried Illicium Verum Hook. f. And Zingiber Officinale Rosc. By improved solvent-free microwave extraction with three types of microwave-absorption medium. Analytical and Bioanalytical Chemistry. Nov. 2006;**386**(6):1863-1868

[18] Michel T, Destandau E, Elfakir C. Evaluation of a simple and promising method for extraction of antioxidants from sea buckthorn (*Hippophaë rhamnoides* L.) berries: Pressurised solvent-free microwave assisted extraction. *Food Chemistry*. 2011;**126**(3):1380-1386

[19] Farhat A, Fabiano-Tixier AS, Visinoni F, Romdhane M, Chemat F. A surprising method for green extraction of essential oil from dry spices: Microwave dry-diffusion and gravity. Journal of Chromatography. A. 2010;**1217**(47):7345-7350

[20] Lucchesi ME, Smadja J, Bradshaw S, Louw W, Chemat F. Solvent free microwave extraction of Elletaria cardamomum L.: A multivariate study of a new technique for the extraction of essential oil. Journal of Food Engineering. Apr. 2007;**79**(3):1079-1086

[21] Qi XL et al. Solvent-free microwave extraction of essential oil from pigeon pea leaves [*Cajanus cajan* (L.) Millsp.] and evaluation of its antimicrobial activity. Industrial Crops and Products. 2014;**58**:322-328

[22] Ma CH, Yang L, Zu YG, Liu TT. Optimization of conditions of solvent-free microwave extraction and study on antioxidant capacity of essential oil from *Schisandra chinensis* (Turcz.) Baill. Food Chemistry. 2012;**134**(4):2532-2539

[23] Uysal B, Sozmen F, Aktas O, Oksal BS, Kose EO. Essential oil composition and antibacterial activity of the grapefruit (*Citrus Paradisi*. L) Peel essential oils obtained by solvent-free microwave extraction: Comparison with hydrodistillation. International Journal of Food Science and Technology. 2011;**46**(7):1455-1461

[24] Riela S et al. Effects of solvent-free microwave extraction on the chemical composition of essential oil of *Calamintha nepeta* (L.) Savi compared with the conventional production method. Journal of Separation Science. 2008;**31**(6-7):1110-1117

[25] Kusuma H, Putri D, Dewi I, Mahfud M. Solvent-free microwave extraction as the useful tool for extraction of edible essential oils. Chemistry & Chemical Technology. 2016;**10**(2):213-218

[26] Li XJ et al. Solvent-free microwave extraction of essential oil from *Dryopteris fragrans* and evaluation of antioxidant activity. Food Chemistry. 2012;**133**(2):437-444

[27] Chen X, Zhang Y, Zu Y, Fu Y, Wang W. Composition and biological activities of the essential oil from *Schisandra chinensis* obtained by solvent-free microwave extraction. LWT- Food Science and Technology. 2011;**44**(10):2047-2052

[28] Paroul N et al. Solvent-free production of bioflavors by enzymatic esterification of citronella (*Cymbopogon winterianus*) essential oil. Applied Biochemistry and Biotechnology. 2012;**166**(1):13-21

[29] Filly A, Fernandez X, Minuti M, Visinoni F, Cravotto G, Chemat F. Solvent-free microwave extraction of essential oil from aromatic herbs: From laboratory to pilot and industrial scale. Food Chemistry. May 2014;**150**:193-198

[30] Li Y, Fabiano-Tixier AS, Vian MA, Chemat F. Solvent-free microwave extraction of bioactive compounds provides a tool for green analytical chemistry. TrAC, Trends in Analytical Chemistry. 2013;**47**:1-11

[31] Okoh OO, Sadimenko AP, Afolayan AJ. Comparative evaluation of the antibacterial activities of the essential oils of *Rosmarinus officinalis* L. obtained by hydrodistillation and solvent free microwave extraction methods. Food Chemistry. 2010;**120**(1):308-312

[32] Périno-Issartier S, Zill-e-Huma, Abert-Vian M, Chemat F. Solvent free microwave-assisted extraction of antioxidants from sea buckthorn (*Hippophae rhamnoides*) food by-products. Food and Bioprocess Technology. 2011;**4**(6):1020-1028

[33] Azar A, Torabbeigi, Sharifan, Tehrani. Chemical composition and antibacterial activity of the essential oil of lavandula angustifolia isolated by solvent free microwave assisted extraction and hydrodistillation. Journal of Food Biosciences and Technology. 2011;**1**:19-24

Section 3

Pharmaceutical Green Chemistry

Chapter 4

Green Chemistry and Synthesis of Anticancer Molecules

Shagufta Perveen and Areej Mohammad Al-Taweel

Additional information is available at the end of the chapter

http://dx.doi.org/10.5772/intechopen.70419

Abstract

Green chemistry is a modern area of chemistry merged with chemical engineering methods. It highlighted the synthesis of molecules in a manner of using environment-friendly chemical reagents with low waste material for enhancing environmental performance which reduce the formation of hazard substances. Modern researches are trying to reduce the risk of human kind health and the environment of our world by doing magnificent work in the field of green chemistry. In the pharmaceutical field, green chemistry works very well with the formation of many drugs and it utilizes non-hazards, reproducible and environment-friendly solvents with low time and money costs by using catalyst, microwave, ultrasonic, solid phase and solvent-free synthesis. Until now, scientist has synthesized many anticancer molecules by using these modern green chemistry techniques. These compounds showed significant anticancer activities against many human cancer cell lines. In this chapter, we will cover different views and the recently published literature to summarize the role of green chemistry in the synthesis of anticancer compounds.

Keywords: green synthetic approaches, anticancer activity, synthesis of active molecules, cancer cell lines

1. Introduction

Green chemistry is a modern way for the synthesis of organic compounds and designed different drugs under facile protocols, efficient conditions, environmentally benign and high yielding method of molecules with advantages over traditional organic synthetic methods. It usually reduces waste by-products, costs and develops environmentally friendly procedures.

Method use	Reaction condition	Class of compounds synthesized	Cancer cell lines	Refs.
Eco-friendly one-pot synthesis	Mixture of methyl ketone, aldehyde, active methylene cyanoacetamide, malononitrile, ethylcyanoacetate and 2 mL of glycol/ammonium acetate was added to the reaction vessel and placed into MW reactor then allowed to react under MW irradiation at 200–400 W power and 120°C for 6–8 min. The compound was collected by filtration and recrystallized from ethanol/DMF to give pure amino-cyanopyridine and oxo-cyanopyridine derivatives	Novel cyanopyridine derivatives (**1**)	Human liver HepG2, colon HCT-116, breast MCF-7 cancer	[1]
Microwave-assisted synthesis	A mixture consisting of methyl salicylate and sodium was heated to 110°C. When the reaction with sodium was completed after 5–10 min, mixture was irradiated for 30 min at 160–200°C using a 200 W MW source. Chromatographic separation of crude mixture on silica gel column gave the pure products	Salicyloyloxy and 2-methoxybenzoyloxy androstane and stigmastane derivatives (**2**)	Human breast adenocarcinoma ER+, MCF-7, estrogen receptor negative breast adenocarcinoma ER+, MDA-MB-231, prostate cancer PC-3 and normal fetal lung fibroblasts MRC-5 cancer	[2]
One-pot synthesis	Reaction of 2-chloro-3-chloromethyl-quinoline with terminal alkyne in the presence of KI, NaN3 and precatalyst copper(II)sulfate in combination with Na-ascorbate was examined in water at room temperature	Quinoline, triazole and dihydroquinoline (**3**)	Human A549 lung adenocarcinoma epithelial, MCF-7 breast adenocarcinoma, HepG2 hepatocellular liver carcinoma, DU145 prostate cancer	[3]
Microwave-assisted synthesis	Ethyl/methyl acetoacetate and an aldehyde were taken into a beaker and dissolved in minimum quantity of dimethylformamide. To this solution, ammonium acetate was added. Reaction mixture was subjected to microwave irradiation at 480 W for 2–6 min, with a pulse rate of 60 s each in a microwave oven. After completion of the reaction on TLC, the resultant product was filtered, washed with chill water and recrystallized	4-alkyl/aryl-3,5-bis(carboethoxy/carbomethoxy)-1,4-dihydro-2,6-imethylpyridines (**4**)	Human HT-29 colon cancer and MDA-MB breast cancer and MRP1 inhibitory activity using the insect cell membrane	[4]

Method use	Reaction condition	Class of compounds synthesized	Cancer cell lines	Refs.
Cellulose-supported copper nanoparticle-catalyzed click reaction in water	4-hydroxybenzaldehyde treated with propargyl bromide in the presence of K2CO3 in dry acetone under reflux to yield 4-*O*-propargylated benzaldehyde. In the next step, 4-*O*-propargylated benzaldehyde was reacted with substituted acetophenones via base-catalyzed Claisen-Schmidt condensation to yield chalcones	Chalcone-linked 1,2,3-triazoles (**5**)	Human MCF-7, MIA-Pa-Ca-2,A549, HepG2 cancer	[5]
Copper-mediated synthesis	Functionalized pyrazolopyridine derivatives via copper-promoted cyclization of pyridyl acetates and benzonitriles in DMSO under argon atmosphere, converted to corresponding pyrazolo[1, 5-a]pyridines from commercially available aromatic nitriles and various pyridyl acetates	Novel pyrazolo, pyridine derivatives (**6**)	Human A549 lung adenocarcinoma, MCF-7 breast carcinoma cell line, HCT-116 colon cancer, PC-3 prostate cancer	[6]
Microwave conditions	Quinolone derivatives were synthesized by reacting 2,3-dihydro-8-nitro-4-quinolones with aromatic aldehydes by pyrrolidine base-catalyzed condensation reaction and were treated with hydrazine derivatives under MW condition, which afforded pyrazolo quinoline derivatives in high yields	Pyrazolo[4,3-c] quinoline (5a-i, 7a-b) and pyrano[3,2-c] quinoline derivatives (**7**)	Human MCF-7 breast and A549 lung cancer	[7]
Facile protocol, efficient and environmentally benign	Synthetic route to barbituric acid derivatives substituted at C5-position. Addition of barbituric acid analogous into nitrostyrene, in water mediated by diethylamine as base gave the target 5-monoalkylbarbiturates in excellent yield	Pyrimidine-2,4,6-trione derivatives (**8**)	HeLa cervical cancer and 3T3 mouse fibroblast cancer	[8]
Simple, eco-friendly and efficient method	To synthesize α,β-unsaturated carbonyl-based compounds, Claisen-Schmidt condensation was used between different ketones and suitable aryl aldehydes in the presence of NaOH in ethanol	α,β-unsaturated carbonyl-based compounds (**9**)	PC12 cancer	[9]

Method use	Reaction condition	Class of compounds synthesized	Cancer cell lines	Refs.
Synthesis by using green solvents	2,3-Dihydrophthalazine-1,4-dione, 5,5-dimethyl clohexane-1,3-dione, aldehyde and p-sulfonic acid calix[4]arene were dissolved in EtOAc. The mixture was irradiated in a MW reactor for 10 min at 130°C. The reaction was cooled to room temperature and then water was added. The mixture was placed in a freezer at 20°C to form the product	Phthalazine-triones: Calix[4]arene (**10**)	Human tumor U251 glioma, MCF7 breast NCIADR/ RES multiple drug-resistant ovarian, 786-0 renal, NCI-H460 lung, non-small cells, PC-3 prostate, OVCAR-03 ovarian, HT-29 colon and K562 leukemia cancer	[10]
One-pot reaction	Mixture of 2-thioxoimidazolidin-4-one and sodium ethoxide in EtOH was refluxed for 30 min. After cooling, CS2 was added and the reaction mixture was stirred at room temperature for 1 h. After evaporation of solution, the solid product was recrystallized from EtOH to give compound in 70–75% yield	Novel 2-thioxoimidazolidin-4-one and benzothiazole thiolate salts (**11**)	MCF-7 breast carcinoma cancer	[11]
One-pot ultrasound-promoted synthesis	One-pot synthesis of 5-amino-2-(4-chlorophenyl)-7-substituted phenyl-8,8a-dihydro-7H-3,4) thiadiazolo(3,2-α)pyrimidine-6-carbonitrile derivatives from three component reactions of 5-(4-chlorophenyl)-1,3,4-thiadiazol-2 amine, aromatic aldehydes and malononitrile in the presence of NaOH under reflux and ultrasonic irradiation	5-amino-2-(4-chlorophenyl)-7-substituted phenyl-8,8a-dihydro-7H-(1,3,4)thiadiazolo (3,2-α)pyrimidine-6-carbonitrile derivatives (**12**)	MCF-7, K562, HeLa and PC-3 cancer	[12]
Microwave-assisted synthesis	Microwave irradiation of mixture of aldehyde and 1,2-phenylenediamine at 80°C, 150 W for 5 min using Na2S2O5 for oxidation, the product 1H-benzo[d] imidazol-2-yl)-6,7,8-trimethoxynaphthalen-1-ol was isolated in excellent 95% yield	2-quinolizinylbenzimidazole and 2-naphthalylbenzimidazole derivatives (**13**)	Human breast MCF-7 cancer	[13]
Laccase-catalyzed green synthesis	These reactions were carried out using catechol, 2,3-dihydro-2-thioxopyrimidin-4(1H)-ones and enzyme laccase, phosphate buffer pH 6 and EtOH with nice yields 95%	Novel pyrimidobenzothiazoles and catechol thioethers (**14**)	Human HepG2 cancer	[14]
Microwave-assisted Hantzsch type condensation reactions	A solution of α-halocarbonyl derivative in dry acetone was added to the solution of benzylidene hydrazine carboseleno amide derivative in DMF. The reaction mixture was stirred at room temperature for 1 day and then neutralized with NaHCO3. The precipitate was filtered and then recrystallized from EtOH	Aryl-hydrazinyl-1,3-selenazole andaroyl-hydrazonyl-1,3-selenazoles (**15**)	Human leukemia cell lines CCRF-CEM and HL60 and carcinoma cell lines MDA-MB231, HCT116 and U87MG	[15]

Method use	Reaction condition	Class of compounds synthesized	Cancer cell lines	Refs.
Catalyst-free, green approach	Mixture of 2-(1H-pyrrol-1-yl) aniline and isatin in EtOH was refluxed at 80°C for 6 h. The progress of reaction was monitored by TLC. On the completion, it cooled to room temperature and then precipitated product was filtered, washed with EtOH and dried by rotavapor to afford pure compound	Pyrrolospirooxindole derivatives (**16**)	Human prostate cancer DU-145	[16]
Simple and convenient one-pot four-component synthesis	Benzaldehyde was reacted with morpholine and 2,4-dinitrophenyl hydrazine in the presence of a chiral pyrrolidine-based catalyst in EtOH. In the next step, compounds were reacted with cinnamaldehyde in the presence of chiral catalyst in toluene at room temperature for 5 h. Compounds were obtained in excellent yields (88–96%)	Morpholine-pyrazolidine derivatives (**17**)	HepG2 liver, HeLa cervical and MCF-7 breast cancer	[17]
Novel one-pot cyclocondensation	A mixture of three components, thiosemicarbazide, 5-acyl thiazoles and phenacyl chlorides, was dissolved in freshly prepared non-volatile organic solvent, DIPEAc and the solution was stirred at room temperature for 30 min. Then, the products were isolated with excellent yields, 82–96%	New bithiazolyl hydrazones (**18**)	MCF-7, HCT116 and THP-1 cancers	[18]
One-pot synthesis	Series of *N*-(aminosulfonyl)-4-podophyllotoxin carbamates were synthesized with amines and *N*-(chlorosulfonyl)-4-podophyllotoxin carbamate dry CH2Cl2 via Burgess-type intermediate, which generated in situ by reaction of PPT and chlorosulfonyl isocyanate CSI in the presence of pyridine	N-(aminosulfonyl)-4-podophyllotoxin carbamates (**19**)	Human tumor HeLa, A-549, HCT-8 and HepG2, human fetal lung fibroblast WI-38 cancer cells	[19]
One-pot solvent-free synthesis	Tryptamine, 2-hydroxy-4,6-dimethylpyrimidine and appropriate thiazole-4-carboxylate were homogenized and then heated at 100–105°C for 5–6 h. The reaction mixture was concentrated and the crude product so obtained was crystallized from EtOH	Bacillamide analogues (**20**)	Human colorectal tumor HCT-116, breast adenocarcinoma MDA-MB-231 and immune system JURKAT cancer	[20]
Microwave-assisted synthesis	Amine and ferulic acid were mixed together in 1:1 ratio for mono-amide and 2:1 molar ratio for bisamide. The reaction mixture was irradiated in microwave at 180–450 Watt for 3–7 min. The reaction progress was monitored by TLC. Products were obtained and purified by crystallization	Ferulic acid amide derivatives (**21**)	Human breast MDA-MB-231 and MCF-7, cervical HeLa, lung A549 and liver HepG2	[21]

Method use	Reaction condition	Class of compounds synthesized	Cancer cell lines	Refs.
Catalyst under solvent-free conditions	The aromatic aldehyde, 2-hydroxy-1,4-naphthoquinone and 2-naphthol grinded in a mortar for 5 min. Then, InCl3 was added and the reaction mixture was grinded 15 min again then placed in a sealed tube and kept in an oven at 120°C for 3 h. The resulting crude was purified by chromatography	Dibenzo anthracenes (**22**)	HEL human erythroleukemia and MCF7 breast cancer	[22]

Table 1. Anticancer molecules by green synthesis.

This chemistry surrounds a series of modern techniques for synthesizing bioactive compounds, such as microwave-assisted synthesis, solid phase supported solvent-free synthesis, reaction with organocatalyst, one-pot multicomponent reactions and sonochemical synthesis, using ionic liquids techniques. Pharmaceutical companies are also improving chemicals to reduce environmental hazards and to minimize ecological risks.

Cancer is a disease generated by uncontrolled cell growth in the body. There are many progresses for cancer treatment but it remains mostly common cause of human death. The number of cancer patients is increasing significantly worldwide, especially in developed countries. According to the global oncology trend report (2015), global spending on cancer medications rose 10.3% in 2014 to $100 billion from $ 75 billion in 2009. Therefore, there is a quick and urgent need of systematic approach to the development of new chemotherapeutic agents with superior efficacy, lower toxicity as well as better selectivity. The methods used in green chemistry organic synthesis of molecules are playing wide role for designing the anticancer drugs. In this chapter, we discuss the most recent literature on green synthesis of different molecules and their anticancer potential on different human cancer cell lines (**Table 1**).

2. Green synthesis of different anticancer molecules

Green chemistry is one of the valuable concepts for the development of new, more effective, solvent-free less toxic, environmentally friendly and cost-efficient methods for the synthesis of different anticancer molecules. There are many developments for the environmentally friendly approaches for the synthesis of biologically active molecules such as microwave-assisted synthesis, one-pot synthesis, solvent-free synthesis, enzyme-catalyzed synthesis, solid phase synthesis, ultrasound promoted and catalyst-free synthesis. Herein, we are discussing some recently published cytotoxic molecules, which have been synthesized by different green synthesis approaches (**Figure 1**).

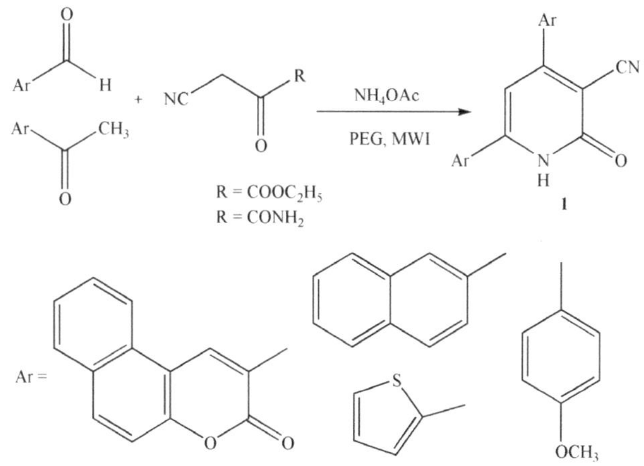

Figure 1. 3-cyano pyridine derivatives.

2.1. One-pot synthesis of 3-cyano pyridine derivatives

Novel series of 3-cyano pyridine type derivatives were synthesized and their cytotoxic activity was evaluated against many human MCF-7, HCT-116 and HepG-2 cancer cell lines. Most of the compounds showed good-to-moderate activity against HepG2 and HCT-116 cell lines, whereas only few compounds showed significant cytotoxic activity against MCF-7 breast cancer cell line (**Figure 1**) [1].

2.2. Microwave-assisted solvent-free synthesis of stigmastane derivatives

The microwave-assisted synthesis in most cases was more successful regarding to the reaction time and the yields of product. These reactions are more environmentally friendly too, compared to the conventional synthetic methods. In this research, a convenient simple microwave-assisted solvent-free synthesis of 2-methoxybenzoyloxy androstane, salicyloyloxy stigmastane derivatives from methyl salicylate and appropriate steroidal precursors has done. 2-Methoxybenzoyl ester exhibited significant cytotoxic activity against MDA-MB-231 cells. Most of the compounds strongly inhibited growth of PC-3 cells, whereas salicyloyloxy stigmastane derivative showed the best inhibition potency (**Figure 2**) [2].

2.3. One-pot synthesis of polyazaheterocycles in water

Synthesis of these polyazaheterocycles was carried out by green synthetic strategy that involved one-pot azidation and CuAAC under mild conditions in water. Many compounds were synthesized and evaluated for their cytotoxic effects against four human cancer cell lines, including A549 (lung), MCF-7 (breast), HepG2 (hepatocellular) and DU145 (prostate). Some of the compounds showed strong activities against A549 cancer cells (**Figure 3**) [3].

2.4. Microwave irradiated one-pot synthesis of carboethoxy/carbomethoxy derivatives

Fourteen carboethoxy/carbomethoxy derivatives have been synthesized by conventional and microwave irradiation method from a one-pot three-component reaction mixture, consisting of, alkyl acetoacetate, aldehyde and ammonium acetate. The synthesized products have been evaluated for their cytotoxic activity against MDA-MB (breast) and HT-29 (colon) human cancer cell lines. Few compounds exhibit some degree of cytotoxicity and it was low when compared with standard (**Figure 4**) [4].

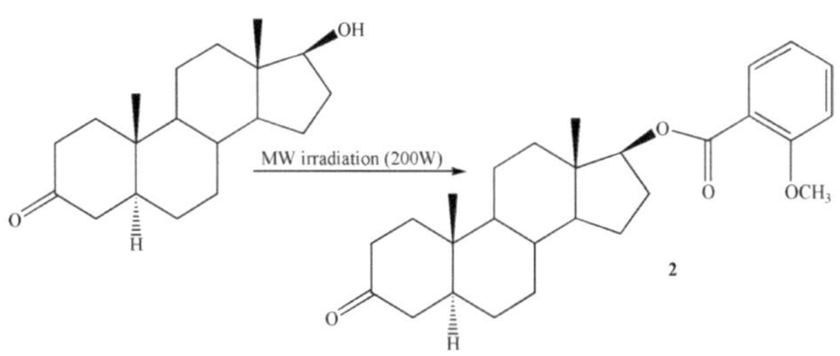

Figure 2. Synthesis of stigmastane derivatives.

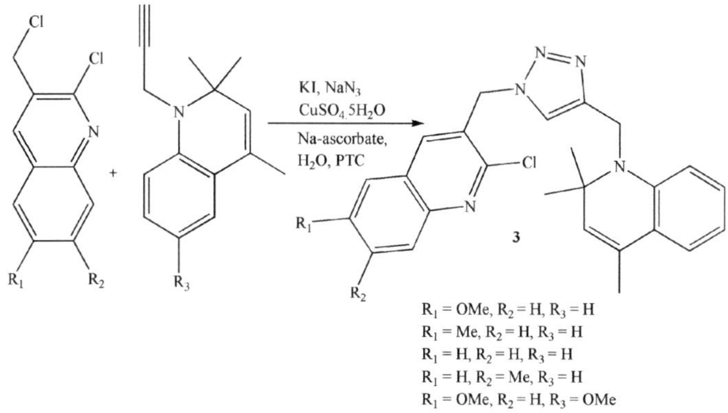

Figure 3. Synthesis of polyazaheterocycles.

2.5. Cellulose-supported copper nanoparticle-catalyzed synthesis of chalcone derivatives

Chalcone-linked 1,2,3-triazole derivatives were synthesized in water by cellulose-supported copper nanoparticle-catalyzed click reaction. All the products were subjected to MTT cytotoxicity assay against four human cancer cell lines A549, MCF-7, HepG2 and MIA-Pa-Ca-2 for testing their anticancer potential. Few compounds were found to be most active against all cancer cell lines and showed better activity when compared to reference drug (**Figure 5**) [5].

2.6. Copper-mediated synthesis of pyrazolo pyridine derivatives

Some novel pyrazolo pyridine type compounds were synthesized by facile procedures and showed significant cytotoxic potential on different human cancer cell lines. They revealed various cancer cell lines (HCT-116, A549, MCF-7, PC-3) determined by SRB assay (**Figure 6**) [6].

2.7. Microwave-assisted synthesis of quinoline analogues

A new class of pyrazolo[4,3-c]quinoline and pyrano[3,2-c]quinoline analogues was synthesized in good yields by microwave conditions. For enhancing the yield of products, multicomponent one-pot synthesis has been developed. The cytotoxicity of these compounds was also evaluated against MCF-7 and A549 cancer cell lines. Most of the compounds displayed moderate-to-good anticancer activity against these cell lines (**Figure 7**) [7].

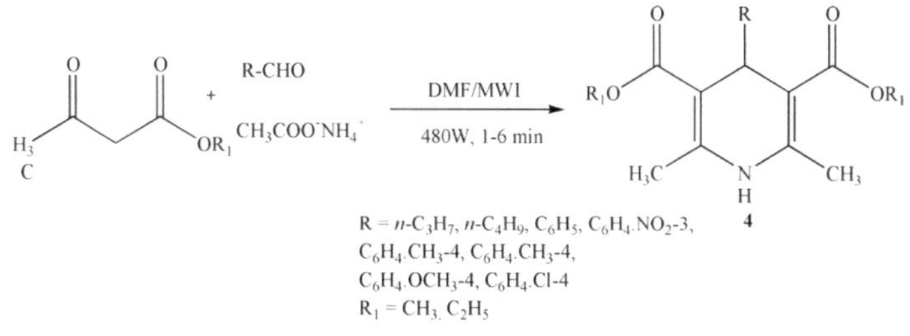

Figure 4. Synthesis of carboethoxy/carbomethoxy derivatives.

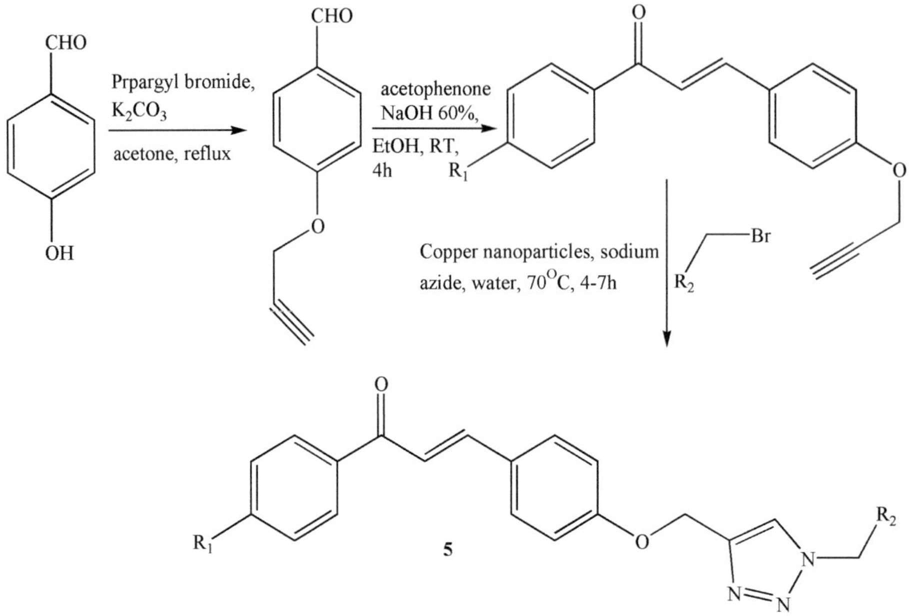

Figure 5. Synthesis of chalcone derivatives.

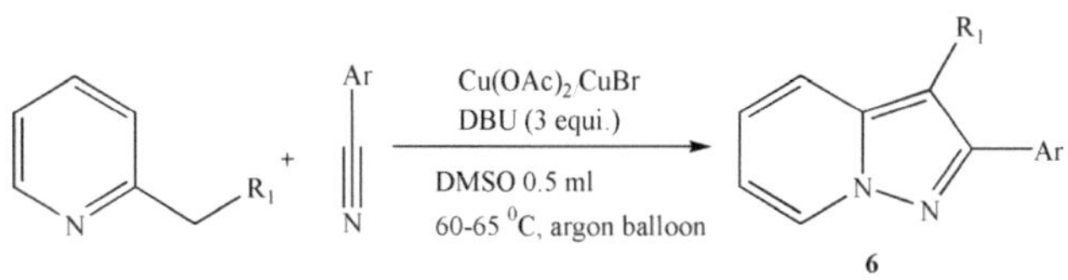

Figure 6. Synthesis of pyridine derivatives.

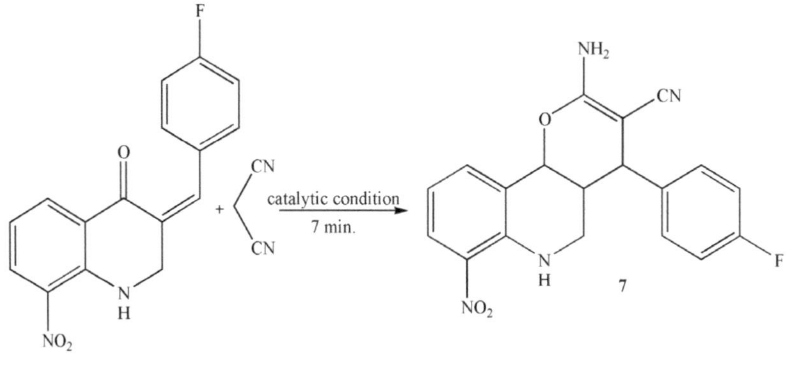

Figure 7. Synthesis of quinoline analogues.

2.8. Facile protocol, efficient and environmentally benign synthesis of cycloheximide

In this research, they describe a facile and efficient protocol and environmentally benign for the synthesis of C5-substituted barbiturate acid in water. The synthesized compounds tested for different assay and provided promising results against a-glucosidase inhibitor. The cytotoxic activity of compound against 3T3 cell resulted that compounds showed significant to weak activity against the standard cycloheximide (**Figure 8**) [8].

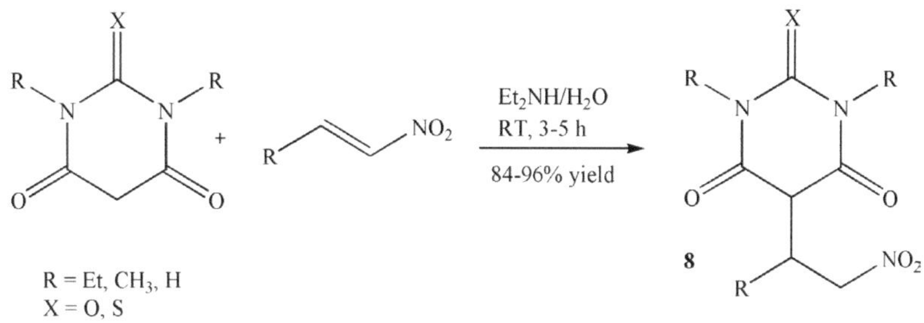

Figure 8. Synthesis of quinoline cycloheximide.

2.9. Simple, eco-friendly and efficient synthesis of α,β-unsaturated carbonyl compounds

A novel series of carbonyl compounds was synthesized by environment-friendly, simple and efficient method. Compounds were tested for cytotoxicity. All strong antioxidant compounds showed strong protective effect against PC12 cell line (**Figure 9**) [9].

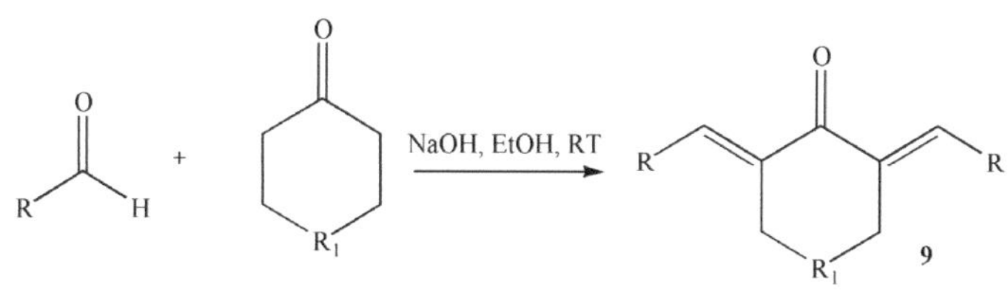

R	R_1
4-Diethoxymethyl-benzene	CH_2
4-Diethoxymethyl-benzene	O
4-Diethoxymethyl-benzene	NH
4-Diethoxymethyl-benzene	NCH_3
4-Diethoxymethyl-benzene	NCH_2-benzene
Phenyl-2-NO_2-4-diethylamine	CH_2
Phenyl-2-NO_2-4-diethylamine	O
Phenyl-2-NO_2-4-diethylamine	NH
Phenyl-2-NO_2-4-diethylamine	NCH_3
Phenyl-2-NO_2-4-diethylamine	NCH_2-benzene
Phenyl-4-pyrrolidine	CH_2
Phenyl-4-pyrrolidine	O
Phenyl-4-pyrrolidine	NH
Phenyl-4-pyrrolidine	NCH_3
Phenyl-4-pyrrolidine	NCH_2-benzene

Figure 9. Synthesis of α,β-unsaturated carbonyl-based compounds.

2.10. Green methodology synthesis of 2H-indazolo[2,1-b]phthalazine-trione derivatives

An efficient green method was used for the synthesis of 2H-indazolo[2,1-b]phthalazine-trione derivatives. Many compounds were obtained in good yields within 10 min. Among all tested cell lines, K562 leukemia cell line was most sensitive (**Figure 10**) [10].

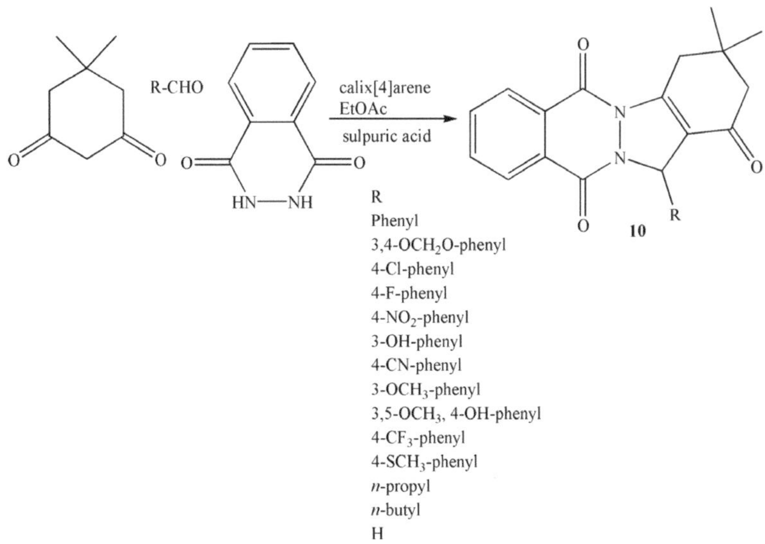

Figure 10. Synthesis of phthalazine-trione derivatives.

2.11. Green synthesis of thioxoimidazolidin and benzothiazole derivatives

A series of 2-thioxoimidazolidin-4-one and benzothiazole thioglycosides were synthesized by one-pot reaction. The cytotoxic activity of compound was evaluated against MCF-7 breast cell and it showed high-to-moderate anticancer activities (**Figure 11**) [11].

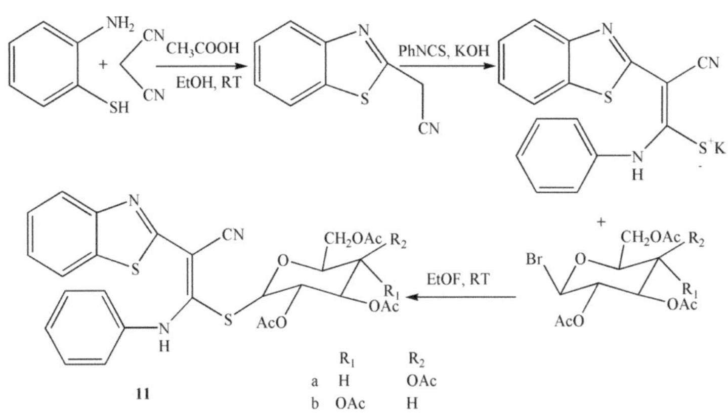

Figure 11. Synthesis of thioxoimidazolidin and benzothiazole derivatives.

2.12. Green synthesis of pyrimidine-6-carbonitrile derivatives

This is a green synthetic approach for the formation of antitumor active 5-amino-2-(4-chlorophenyl)-7-substituted phenyl-8,8a-dihydro-7H-(1,3,4)thiadiazolo(3,2-α) pyrimidine-6-carbonitrile. This protocol is extendable to a wide variety of many substrates. The advantages are the use of eco-friendly catalyst, reduced time, simple work-up process, ease of isolation and high yield of product. One compound was found to have the highest GI50 value for PC-3, HeLa, K562 and MCF-7 cancer cell lines (**Figure 12**) [12].

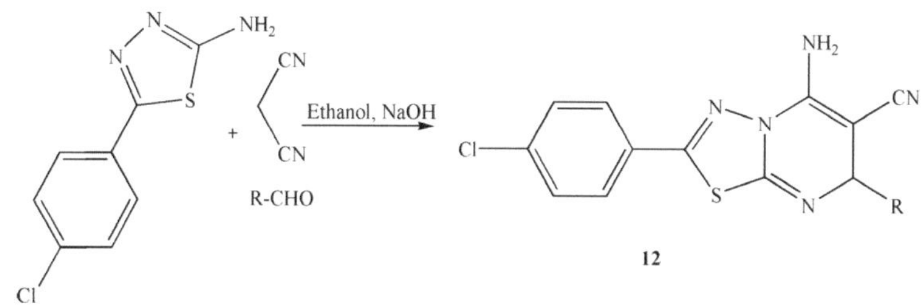

Figure 12. Synthesis of pyrimidine-6-carbonitrile derivatives.

2.13. Microwave-assisted synthesis of benzimidazole derivatives

Twelve 2-quinolizinylbenzimidazole and 2-naphthalylbenzimidazole type compounds have been synthesized under MW microwave condition. These compounds were tested for cytotoxicity against human breast cancer cell line MCF-7. The results showed that some compounds were found to be as active as standard Tamoxifen (**Figure 13**) [13].

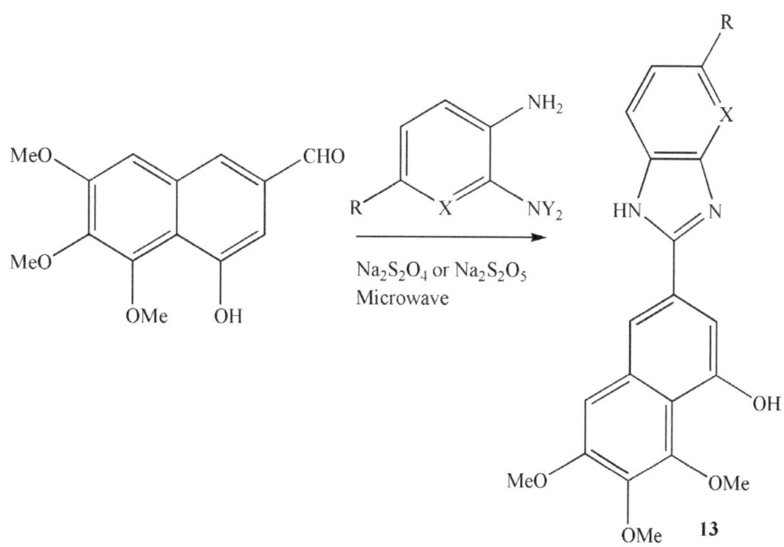

Figure 13. Synthesis of pyrimidine-6-carbonitrile derivatives.

2.14. Enzyme laccase-catalyzed green synthesis of pyrimidobenzothiazoles

This is a newly developed method for the synthesis of pyrimidobenzothiazoles and catechol thioethers, and it addressed many of the principles of green chemistry. These reactions were catalyzed by laccase enzyme and transformations were completely safe and non-toxic aerial oxygen as the sole oxidant. This reaction delivers the products in an excellent yield. Among all tested compounds, few compounds showed moderate-to-good activity against HepG2 cell line (**Figure 14**) [14].

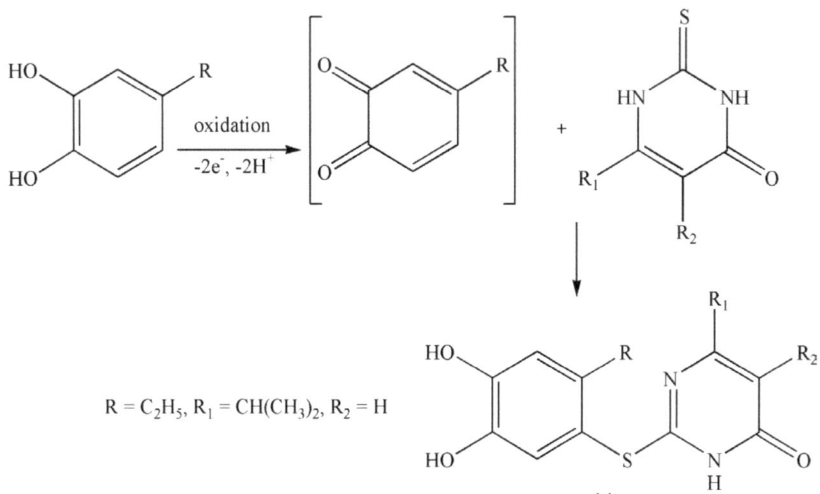

14

Figure 14. Synthesis of pyrimidobenzothiazoles.

2.15. Microwave-assisted synthesis of 1,3-selenazole derivatives

Synthesis of new 1,3-selenazole derivatives has been done by MW-assisted Hantzsch condensation reactions. Compound were screened for anti-proliferative effects against leukemia cell lines (HL60 and CCRF-CEM) and carcinoma cell lines (HCT116, MDA-MB231 and U87MG) and it gave moderate cytotoxicity against all tested cell lines (**Figure 15**) [15].

2.16. Environment-friendly synthesis of 5′H-spiro[indoline-3,4′-pyrrolo(1,2-a) quinoxalin]-2-ones

A very simple-to-perform, efficient, mild and environment-friendly benign formation of 5′H-spiro[indoline-3,4′-pyrrolo(1,2-a)quinoxalin]-2-ones has been developed without any catalysts. This method includes simplicity of operation, clean reaction, no side products and good yields. Purification of product is very simple, involving a filtration and washing. The synthesized compound with piperonyl substitution on 5-chloroisatin nitrogen showed highest cytotoxicity. Its IC_{50} values are comparable to that of the standard doxorubicin (**Figure 16**) [16].

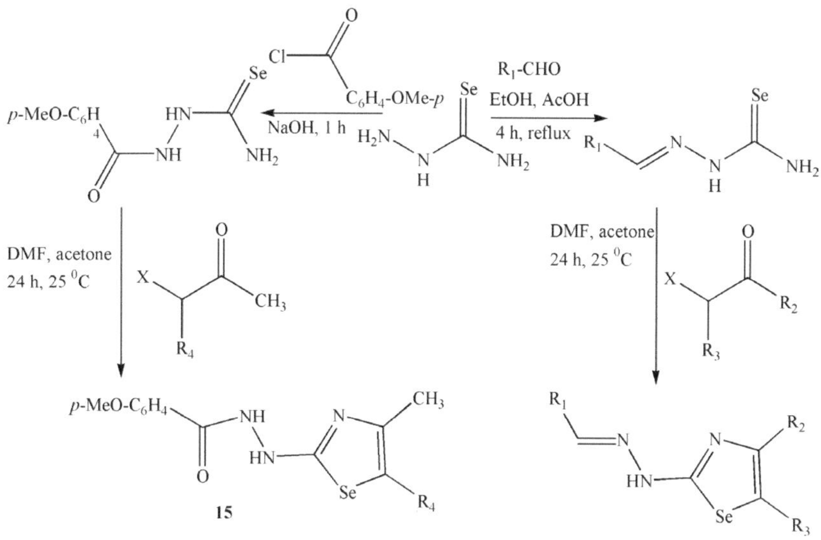

Figure 15. Synthesis of 1,3-selenazole derivatives.

2.17. One-pot four-component synthesis of morpholine-connected pyrazolidine derivatives

A simple and convenient one-pot four component reaction of morpholine connected with pyrazolidine derivatives was developed using metal-free catalysis. Cytotoxicity was evaluated using HeLa (cervical), HepG2 (liver) and MCF-7 (breast) cancer cell lines, and compounds showed significant cytotoxicity against tested cells (**Figure 17**) [17].

2.18. Novel one-pot cyclocondensation

Bithiazolyl hydrazones have been synthesized by one-pot cyclocondensation reaction in freshly prepared ionic liquid at room temperature. Compounds have been evaluated for antitubercular activity and showed potent antitubercular activity (**Figure 18**) [18].

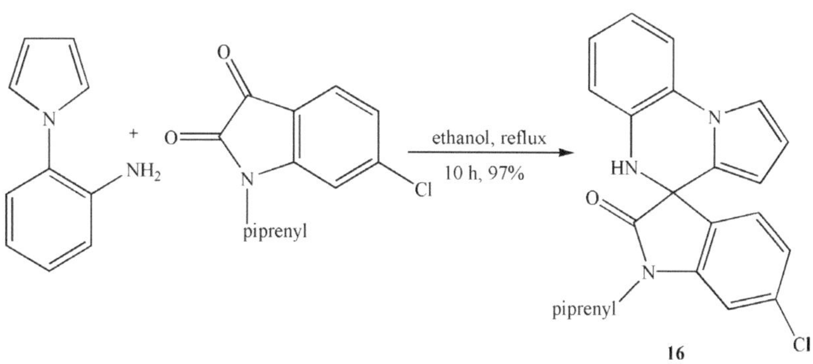

Figure 16. Synthesis of spiro[indoline-3,4′-pyrrolo(1,2-a)quinoxalin.

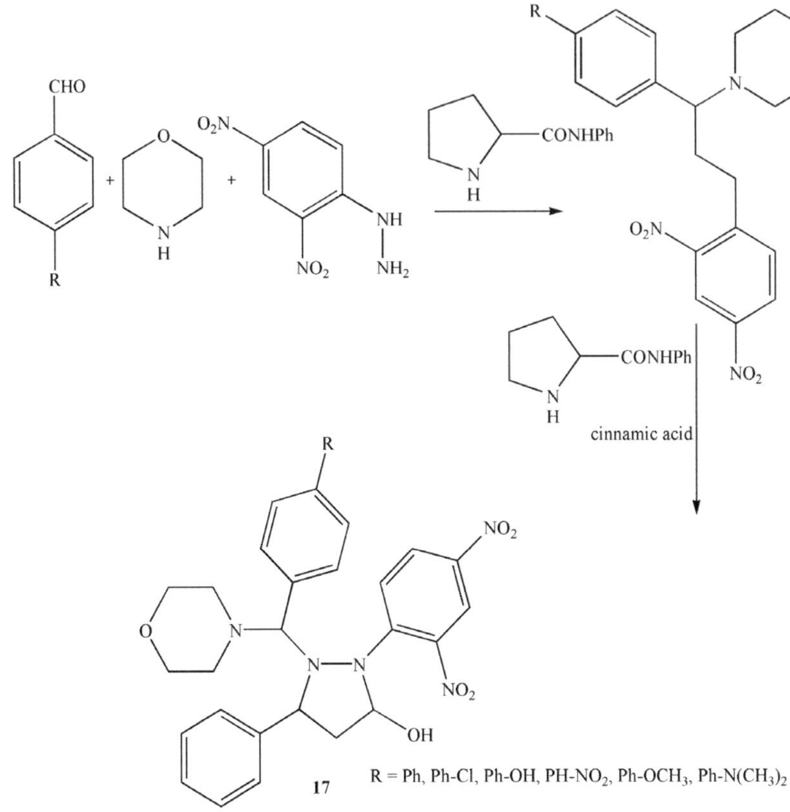

Figure 17. Synthesis of morpholine-connected pyrazolidine.

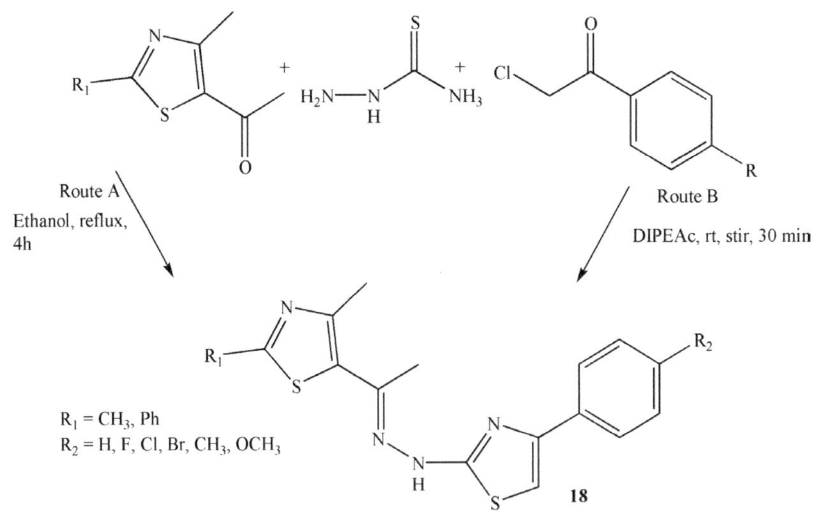

Figure 18. Synthesis of bithiazolyl hydrazones.

2.19. One-pot synthesis of carbamates

One-pot synthesis of N-(aminosulfonyl)-4-podophyllotoxin carbamates has been done and it showed promising cytotoxic activities. Most effective compound induced HeLa cells cycle arrest in G2/M phase, leading to apoptosis, and activation of cdc2, cyclinB1, p53 and ROS and inhibits polymerization of tubulin and microtubule. These results suggest that these synthesized compounds have strong potential for development as cytotoxic agents (**Figure 19**) [19].

2.20. Eco-friendly synthesis of bacillamide

It is an efficient, advanced and eco-friendly route for synthesis of bacillamide analogues through a two-step solvent-free synthesis. Compounds exhibit potent cytotoxic activity against three HCT-116, MDA-MD-231 and JURKATs cancer cell lines and compared with doxorubicin (**Figure 20**) [20].

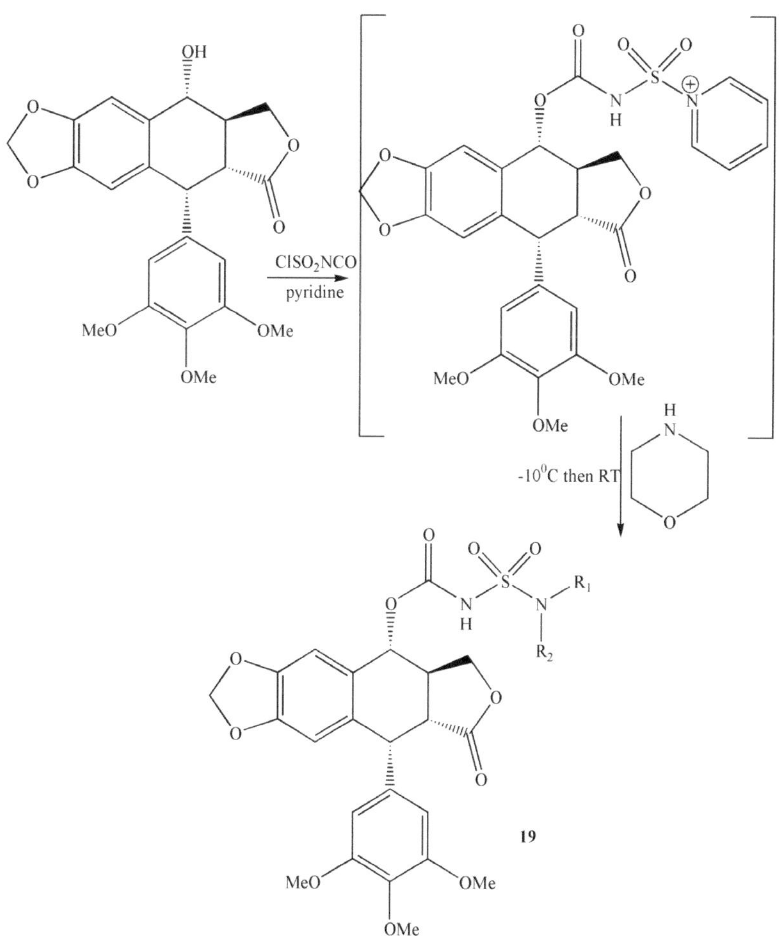

Figure 19. Synthesis of N-(aminosulfonyl)-4-podophyllotoxin carbamates.

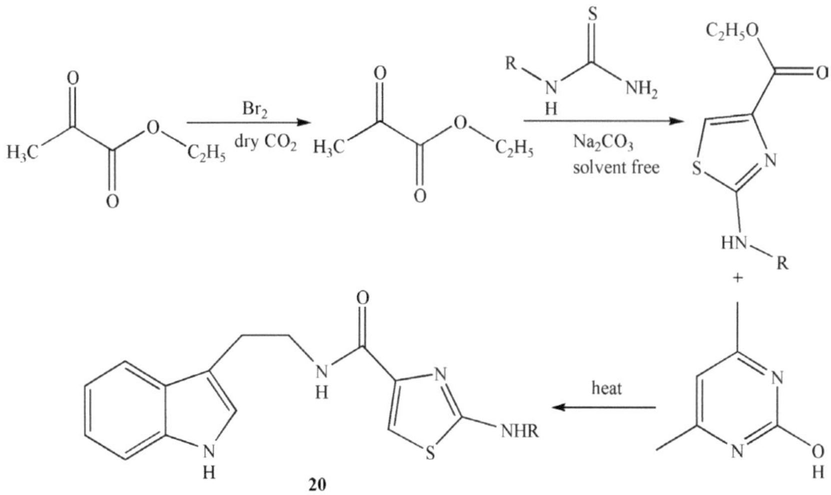

Figure 20. Synthesis of bacillamide analogues.

2.21. Solvent-free microwave-assisted synthesis of amide derivatives of ferulic acid

In this research work, different amide derivatives of ferulic acid have been synthesized under solvent-free conditions by microwave-assisted reaction. These compounds were found to exhibit noticeable in vitro anticancer activity against breast (MDA-MB-231 and MCF-7), cervical (HeLa), lung (A549) and liver (HepG2) human cancer cell lines (**Figure 21**) [21].

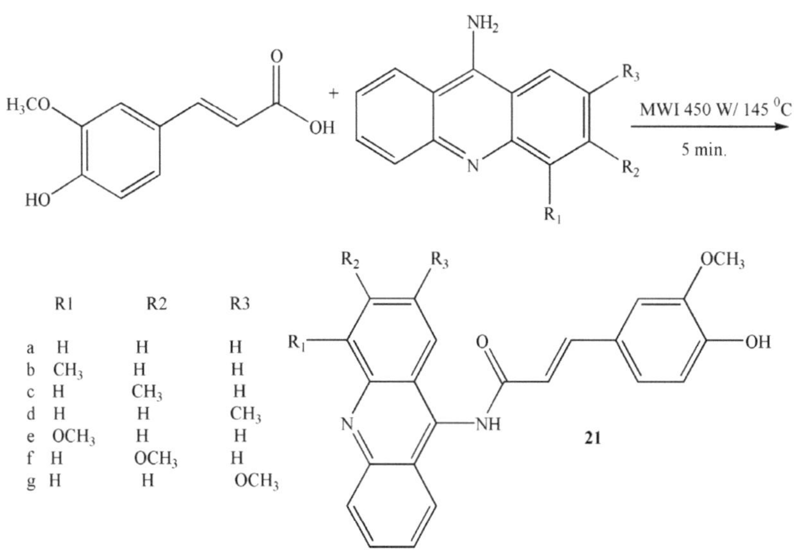

	R1	R2	R3
a	H	H	H
b	CH_3	H	H
c	H	CH_3	H
d	H	H	CH_3
e	OCH_3	H	H
f	H	OCH_3	H
g	H	H	OCH_3

Figure 21. Synthesis of ferulic acid derivatives.

2.22. One-pot synthesis of *o*-quinonic adducts

Dibenzo[a,h]anthracene derivatives were synthesized via a one-pot synthetic protocol with three-component reaction of 2-hydroxy-1,4-naphthoquinone, aromatic aldehydes and 2-naphthol using InCl3 as catalyst under solvent-free condition. These *o*-quinonic adducts showed strong cytotoxicity against MCF-7 and HEL tumoral cell lines (**Figure 22**) [22].

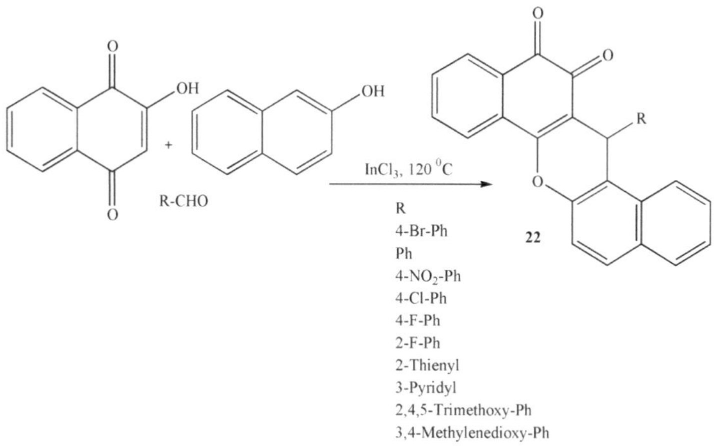

Figure 22. Synthesis of *o*-quinonic adducts.

3. Conclusion

The data of this chapter could be very helpful to identify the recently published approaches of anticancer molecules synthesized via different green chemistry approaches.

Acknowledgements

This book chapter was supported by a grant from the "Research Center of the Female Scientific and Medical Colleges," Deanship of Scientific Research, King Saud University.

Author details

Shagufta Perveen* and Areej Mohammad Al-Taweel

*Address all correspondence to: shagufta792000@yahoo.com

Department of Pharmacognosy, College of Pharmacy, King Saud University, Riyadh, Saudi Arabia

References

[1] Khaled AMA, Ghada HA, Abeer ME. Eco-friendly synthesis of novel cyanopyridine derivatives and their anticancer and PIM-1 kinase inhibitory activities. European Journal of Medicinal Chemistry 2017;**134**:357-365. DOI: 10.1016/j.ejmech.2017.04.024.

[2] Katarina MPG, Evgenija AD, Mihaly S, Janos G, Janos JC. Microwave assisted synthesis and biomedical potency of salicyloyloxy and 2-methoxybenzoyloxy androstane and stigmastane derivatives. Steroids 2015;**94**:31-40. DOI: 10.1016/j.steroids.2014.12.008

[3] Koduru SSP, Edupuganti VVSR. Design of new hybrid template by linking quinoline, triazole and dihydroquinoline pharmacophoric groups: A greener approach to novel polyazaheterocycles as cytotoxic agents. Bioorganic & Medicinal Chemistry Letters 2015;**25**:1057-1063. DOI: 10.1016/j.bmcl.2015.01.012

[4] Srinivas NA, Mahendar P, Sadanandam A, Achaiah G, Malla RV. Synthesis, anticancer and MRP1 inhibitory activities of 4-alkyl/aryl-3,5-bis(carboethoxy/carbomethoxy)-1, 4-dihydro-2, 6-dimethylpyridines. Medicinal Chemistry Research 2013;**22**:147-155. DOI: 10.1007/s00044-012-9994-0

[5] Pinki Y, Kashmiri L, Ashwani K, Santosh KG, Sundeep J. Green synthesis and anticancer potential of chalcone linked-1,2,3-triazoles. European Journal of Medicinal Chemistry 2017;**126**:944-953. DOI: 10.1016/j.ejmech.2016.11.030

[6] Chitrakar R, Arem Q, Darapaneni CM, Shashank KS. Design, synthesis and cytotoxicity studies of novel pyrazolo[1, 5-a] pyridine derivatives. European Journal of Medicinal Chemistry 2017;**126**:277-285. DOI: 10.1016/j.ejmech.2016.11.037

[7] Thangaraj A, Sadasivam M, Subashini G, Selvaraj S, Athar A, Palathurai SM. Biologically active perspective synthesis of heteroannulated 8-nitroquinolines with green chemistry approach. Bioorganic & Medicinal Chemistry Letters 2017;**27**:1538-1546. DOI: 10.1016/j.bmcl.2017.02.042

[8] Assem B, Mohammad SI, Abdullah MA, Hazem AG, Sammer Y, Mahwish A, Nimra NS, Choudhary MI, Ruqaiya K, Zaheer U. Synthesis of pyrimidine-2,4,6-trione derivatives: Anti-oxidant, anti-cancer, a-glucosidase, b-glucuronidase inhibition and their molecular docking studies. Bioorganic Chemistry 2016;**68**:72-79. DOI: 10.1016/j.bioorg.2016.07.009

[9] Syed NAB, Ibrahim J, Vijay HM, Devidas TM, Muhammad S, Hassan MN, Muhammad WA. Synthesis of α, β-unsaturated carbonyl based compounds as acetylcholinesterase and butyrylcholinesterase inhibitors: Characterization, molecular modeling, QSAR studies and effect against amyloid β-induced cytotoxicity. European Journal of Medicinal Chemistry 2014;**83**:355-365. DOI: 10.1016/j.ejmech.2014.06.034

[10] Yuri FR, Cleiton MS, Daniel LS, Jeferson GS, Ana LTGR, Joao EC, Sergio AF, Angelo F. Phthalazine-triones: Calix[4]arene-assisted synthesis using green solvents and their anticancer activities against human cancer cells, Arabian Journal of Chemistry (2016) in press. DOI: 10.1016/j.arabjc.2016.04.007

[11] Elgemeie GH, Farag AB, Amin KM, El-Badry OM and Hassan GS. Design, synthesis and cytotoxic evaluation of novel heterocyclic thioglycosides. Medicinal Chemistry 2014;**4**:814-820. DOI: 10.4172/2161-0444.1000234

[12] Shailee VT, Julio AS, Vazquez-Tato MP, Aniket PS, Deepak KL, Anna PGN. Ultrasound mediated one-pot, three component synthesis, docking and ADME prediction of novel 5-amino-2-(4-chlorophenyl)-7-substituted phenyl-8,8a-dihydro-7H-(1,3,4)thiadiazolo (3, 2-α) pyrimidine-6-carbonitrile derivatives as anticancer agents. Molecules 2016;**21**:894. DOI: 10.3390/molecules21080894

[13] Hue TBB, Quy TKH, Won KO, Duy DV, Yen NTC, Cuc TKT, Em CP, Phuong TT, Loan TT, Hieu VM. Microwave assisted synthesis and cytotoxic activity evaluations of new benzimidazole derivatives. Tetrahedron Letters 2016;**57**:887-891. DOI: 10.1016/j.tetlet.2016.01.042

[14] Abdel-Mohsen HT, Conrad J, Harms K, Nohr D, Beifuss U. Laccase-catalyzed green synthesis and cytotoxic activity of novel pyrimidobenzothiazoles and catechol thioethers. RSC Advances 2017; **7**:17427-17441. DOI: 10.1039/c6ra28102h

[15] Adriana IG, Luiza G, Victor K, Luminita S, Thomas E, Valentin Z. Microwave-assisted synthesis of new selenazole derivatives with antiproliferative activity. Molecules 2013;**18**: 4679-4688. DOI: 10.3390/molecules18044679

[16] Ahmed K, Rasala M, Nayak VL, Korrapati SB, Kumar GB, Anver BS, Jeevak SK, Abdullah A. Discovery of pyrrolospirooxindole derivatives as novel cyclin dependent kinase 4 (CDK4) inhibitors by catalyst-free, green approach. European Journal of Medicinal Chemistry 2016;**108**:476-485. DOI: 10.1016/j.ejmech.2015.11.046

[17] Radhakrishnan SK, Meera M, Salem SA, Aseer M, Akbar I. Synthesis of new morpholine-connected pyrazolidine derivatives and their antimicrobial, antioxidant, and cytotoxic activities. Bioorganic & Medicinal Chemistry Letters 2017;**27**:66-71. DOI: 10.1016/j.bmcl.2016.11.032

[18] Mahendra BB, Sambhaji TD, Amarsinh RD, Laxman UN, Vijay K, Dhiman S, Ramrao AM. New bithiazolyl hydrazones: Novel synthesis, characterization and antitubercular evaluation. Bioorganic & Medicinal Chemistry Letters 2017;**27**:288-294. DOI: 10.1016/j.bmcl.2016.11.056

[19] Xiao-Hui X, Xiao-Wen G, Shi-Liang F, You-Zhen M, Shi-Wu C, Ling H. One-pot synthesis and biological evaluation of N-(aminosulfonyl)-4-podophyllotoxin carbamates as potential anticancer agents. Bioorganic & Medicinal Chemistry Letters 2017;**27**:2890-2894. DOI: 10.1016/j.bmcl.2017.04.082

[20] Sunil K, Ranjana A, Virender K, Rachna S, Bhumi P, Pawan K, Dhirender K. Solvent-free synthesis of bacillamide analogues as novel cytotoxic and anti-inflammatory agents. European Journal of Medicinal Chemistry 2016;**123**:718-726. DOI: 10.1016/j.ejmech.2016.07.033

[21] Naresh K, Sandeep K, Sheenu A, Kumar N, Sham MS, Prasad VB, Partha R, Vikas P. Ferulic acid amide derivatives as anticancer and antioxidant agents: Synthesis, thermal,

biological and computational studies. Medicinal Chemistry Research 2016;**25**:1175-1192. DOI: 10.1007/s00044-016-1562-6

[22] Idaira HF, Angel A, Patricia M, Matias L, Leandro F, Ana E. Indium catalyzed solvent-free multicomponent synthesis of cytotoxic dibenzo[*a,h*]anthracenes from aldehydes, 2-hydroxy-1,4-naphthoquinone, and 2-naphthol. Tetrahedron 2014;**70**:8480-8487. DOI: 10.1016/j.tet.2014.09.076

Chapter 5

The Role of Green Solvents and Catalysts at the Future of Drug Design and of Synthesis

Nurettin Menges

Additional information is available at the end of the chapter

http://dx.doi.org/10.5772/intechopen.71018

Abstract

Green chemistry is getting extended in many researches and industry areas. Not only pharmaceutical companies but also the other chemical industries started to take a step for green chemistry due to its advantages such as decreasing of waste and cost. With this respect, we have already witnessed that pharmaceutical companies searched out for green protocol when manufactured the pharmaceuticals. Green chemistry strategies can be seen in solvents, catalysts, and the others. So, we have briefly discussed the green solvents and nanocatalysts in this chapter. We hope that this chapter gives a brief consideration of importance of green chemistry.

Keywords: nanocatalyst, pharmaceutical company, green chemistry, environment

1. Introduction

The demand of green chemistry for applying in the pharmaceutical and the other chemical industries is increasingly vital due to the fact that our world faces the environmental challenges of the twenty-first century. US Environmental Protection Agency (EPA) has suggested green chemistry for innovative technologies that reduce toxic, undesired waste, and environmental impact. Green chemistry is thus getting grew as an open light to afford a huge scientific area. After EPA, 12 principles of green chemistry have been gotten more attention and these principles have been considered more seriously by pharmaceutical companies since 1998. Pharmaceutical companies declared that they should improve the environmental performance by utilizing green chemistry. Not only pharmaceutical companies but also the other chemical industries started to take a step for green chemistry due to its advantages such as decreasing of waste and cost. It is assumed that green chemistry can save the industry an estimated USD 65.5 billion by 2020 [1] primarily by reducing manufacturing costs. If the processes can be implemented right, green chemistry can afford to reduce waste product and decrease the resources consumption. Green chemistry is needed for minimizing of some social risks and safety issues, as well.

In the past decade, some of the large pharmaceutical companies around the world have focused on using green chemistry processes for drug discovery, development, and manufacturing. These firms include Amgen, the Merck Group, Abbott, Eli Lily, Johnson & Johnson, and Roche [2]. Green chemistry has started to point three lines such as cost, mankind, and our planet. American Chemical Society's Green Chemistry Institute's Pharmaceutical Roundtable was therefore launched and, since 2008, many drug companies have become the members that aim to foster the development of more efficient, less polluting processes. Fortunately, green chemistry celebrates 25 years of progress on 2016 [3].

It can not be denied that people need medicines to cure their diseases some of which are very unpleasant. For that reason, pharmaceutical industry has struggled to have modern synthetic strategies for known and unknown therapeutic reagents. On the other hand, although many successful methodologies were achieved, the toxic properties of many reagents and solvents were not known and the issues of waste minimization and sustainability of solvents and/or unreacted reagents were largely unheard. Chemists and medicinal scientists can reduce the risk to human health and the environment by following all the valuable principles of green chemistry. The most simple and direct way to apply green chemistry in pharmaceuticals is to utilize eco-friendly, nonhazardous, reproducible, and efficient solvents and catalysts in the synthesis of drug molecules, and in researches involving synthetic chemistry.

It has become clear that the chemical and related industries such as pharmaceutical companies are faced with environmental problems. There are a lots of synthetic methodologies and they have generated abundant amounts of waste and chemical industries want to minimize or eliminate this waste. Sheldon has discussed that in the pharmaceutical industries, there is an urgency for consideration of the waste product as a number of by-products produced per kg of product (designated E factor) is about 25 [4]. Innovative strategies on chemistry are the core of the pharmaceutical business. The main point is gathering technology and chemistry to improve lives of patients and minimize environmental impact.

Solvents and stoichiometric reagents are the most important parameters to be considered for greener strategies and these parameters are under investigation in detail by many pharmaceutical companies such as Sanofi-Aventis [5, 6] and GlaxoSmithKline [7]. These companies have suggested that conventional solvents such as halogenated, petroleum-based should be converted into greener solvents such as glycerol, ethyl lactate [8], and water [9]. A catalyst is an another crucial parameter which reduce the amount of inorganic salts and/or reagents. Green alternative for consuming of stoichiometric salts and reagents is to use a catalyst and this issue has been considered by pharmaceutical companies. However, demanding of the least expensive reagents has limited the applying of catalysts to be used widely.

Future perspective of green chemistry will be extended more seriously in many research areas. Product and environment should be considered together and it should be remembered that this planet needs a balance of nature. Every attempt to heart this balance will come across more serious effects. That is why we need greener strategies and greener thinking. In this chapter, we have discussed the importance of solvents and catalysts for synthetic strategy at pharmaceutical chemistry. The progress and advantages about green solvents and bio- and organic-catalysts will be included in detail and we hope that whole of this knowledge will be a hand for both medicinal scientists and pharmaceutical industries.

2. Synthetic strategies with green solvents

In every product development processes and different industrial applications, solvents are needed in huge amounts resulting in abundant amounts of waste. Innovative technologies and different synthetic strategies have discussed solventless methods which are not accepted for all areas of research due to some market concerns. After the solventless ideas, chemists and medicinal scientists have searched out for solvents which suit green chemistry. According to Fischer, green solvent expresses the target to minimize the environmental impact coming from the consuming of solvents in chemical production [9]. Some strategies have emerged for solvents which can be mentioned as green. These are substitution of hazardous solvents resulting in more eco-friendly, biodegradable, and/or minimizing of ozone depletion potential, use of biosolvents (oleochemicals), and substitution of organic solvents which are supercritical fluids and ionic liquids [10].

In the literature, many examples of green solvents can be seen for forming natural products, medicines, and important intermediate products which can be used for further synthesis.

2.1. Vegetable oils as a green solvent

Vegetable oils are oleochemicals which are extracted from many plants' seeds. They are renewable resources and have triglyceride structure in which three hydroxyl groups of glycerol are substituted with different fatty acids that make them liquids or solid products [11]. Vegetable oils are important food ingredient. Unfortunately, they have not considered as a green solvent so far except for a reaction which was published by us [12]. Vegetable oils have been utilized for biopolymers and might be evaluated by scientists who are looking for a new source of green solvent.

We have described the acylation and cyclization reaction that has been run in vegetable oil, especially corn oil. Utilization and advantages of vegetable oils have been discussed and yields, reaction times, and sustainability of vegetable oils have been compared both with each other and with toxic solvent, xylene. This reaction is the first example of vegetable oils and this idea should be concerned by more synthetic strategies due to the cost and efficacy of vegetable oils (**Scheme 1**) [12].

A mixture of dibenzoylmethane (**1**), oxalyl chloride (**2**), and phenol (**3**) was heated in corn oil at 120°C for 15 min. Authors have explained that CH_2 of compound **1** was acylated very easily.

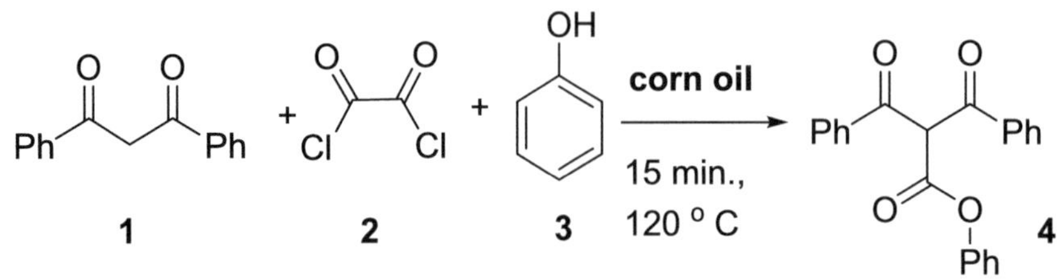

Scheme 1. Acylation reaction in corn oil [12].

2.2. Glycerol as a green solvent

Glycerol (also named as glycerin) is a polyalcohol and second part of oleochemicals which are derived from natural oils. Glycerol has been utilized in many different fields such as pharmaceutical and food industry, tobacco, and cellulose films [13]. Sustainability and low-cost of glycerol make it a good green solvent. With this respect, pharmaceutical companies and chemists have gotten more attention for glycerol as alternative to other organic solvents which are hazardous, volatile compounds, toxic, and harmful. Despite the fact that glycerol is a solvent and selected for many reactions, there are some obstacles which chemists and medical scientists have to overcome: (i) due to the viscosity of glycerol, it should be fluidified with a co-solvent. On the other hand, glycerol is much less viscous up to 60°C and reactions can be proceeded at temperatures higher than 60°C; (ii) glycerol might join the reaction as a reagent, as it has three OH groups which can be mentioned as acidic sites; (iii) glycerol has an enough length and donor atom in which it can obtain complexes with metal catalysts resulting in unwanted side products and/or unreactivity of catalysis. It can be said that there are two sides of glycerol and those can be mentioned for every solvents and reagents which are used in research areas. However, in here, we want to display advantages of glycerol in synthetic strategies.

Safaei et al. synthesized *4H*-pyrans with catalyst-free, one-pot and three-component strategy using glycerol as green solvent (**Scheme 2**) [14]. Yield of reactions are high up to 93% and reactions gave many different types of pyran derivatives. Furthermore, authors have tested the reaction in water and they have seen that yield of the reactions was decreased down to 70%.

Cyclization reaction under atom economic and green solvent procedure is so important and this kind of reactions has prompted medicine scientists to reorganize the strategy for drug design.

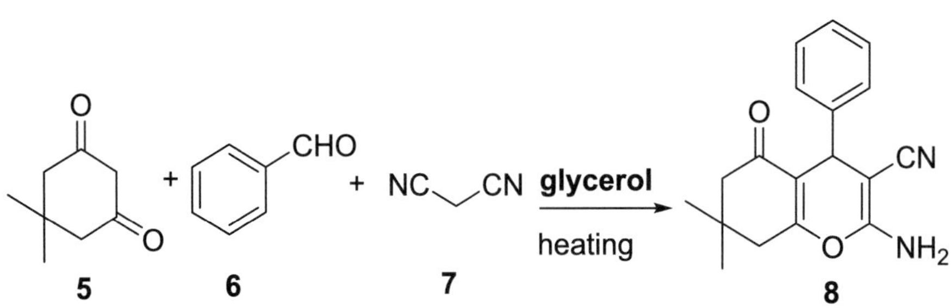

Scheme 2. One-pot and three-component strategy in glycerol [14].

http://dx.doi.org/10.5772/intechopen.71018

Jerome et al. have reported some common synthetic reactions on glycerol and they have discussed yield, reaction time, and sustainability of glycerol. One of the reactions that researchers have carried out was a nucleophilic attack on the β position of α,β-unsaturated carbonyl molecule. This type of reaction can be seen in many organic reactions and derivatization of lead compounds in pharmaceutical industries. The reaction showed that crude glycerol gave yield of the reaction up to 81% and reuse of glycerol as third time for same reactions did give the yield of the product perfectly (**Scheme 3**) [15].

Selectivity and green perspective of the reaction were evaluated by researchers. It is obvious that the reaction aforementioned was progressed in high yield on glycerol and the other organic solvents such as toluene, DMF, and DMSO did not produce any amount of expected product.

Glycerol is used for different reactions and demands on different application fields can be surpassed with modification of glycerol with simple thinking. This modification was done by Garcia et al. [16]. Garcia and his group have synthesized some alkylated derivatives of glycerol to be used as a solvent and evaluated their physical and chemical properties for further applications.

Organochalcogens are under investigation because of the importance of organochalcogen which includes one of an atom of Group 16 in the periodic table which are O, S, Se, and Te. Ebselen and its analogs are important molecules which show significant beneficial effects in primate model of neurodegenerative diseases (**Figure 1**) [17–19]. Organoselenium compound **12** is currently in clinical trials for cardiovascular indications [17]. The importance of these molecules has opened an area for the furnishing of these types of molecules.

With this respect, Leonardo et al. have described a green protocol without base and metal in glycerol for obtaining of organoselenium derivatives (**Scheme 4**) [19].

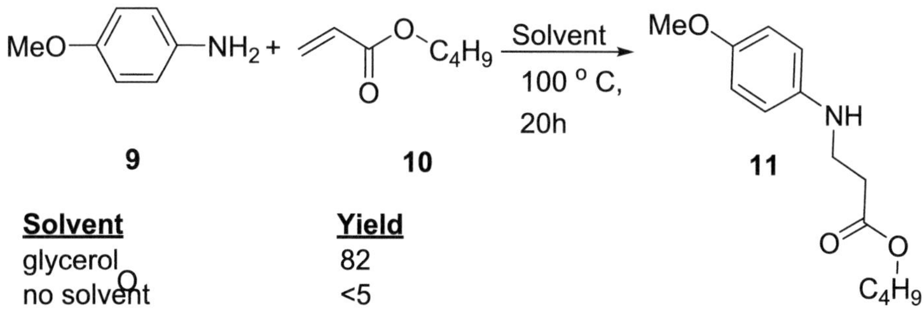

Solvent	Yield
glycerol	82
no solvent	<5
toluene	0
DMF	0
DMSO	0
water	<5
1,2-propandiole	30

Scheme 3. Nucleophilic attack to the α,β-unsaturated carbonyl group in glycerol [15].

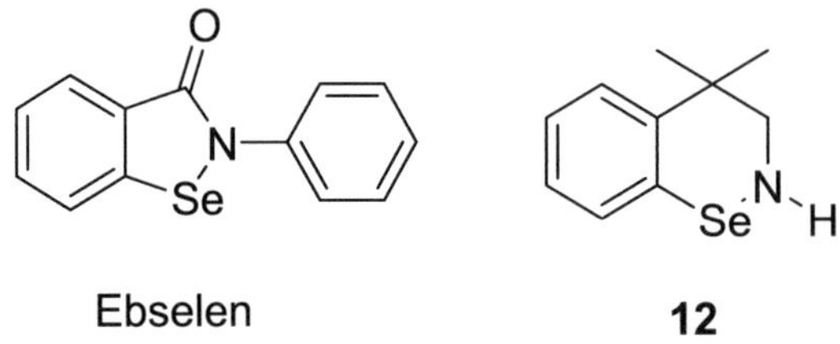

Figure 1. Two examples of organochalcogens.

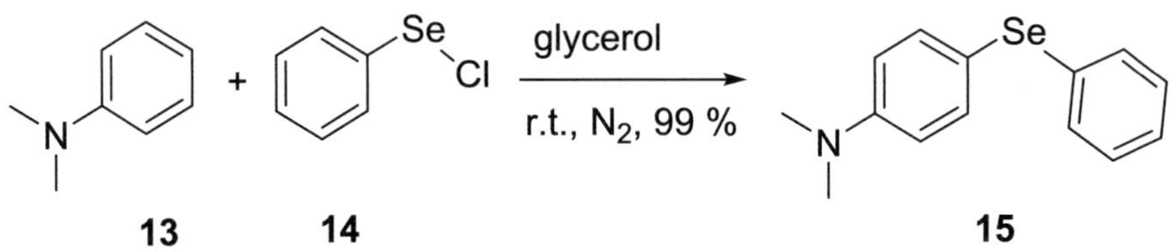

Scheme 4. Synthesis of organoselenium derivative in glycerol [19].

N,N-dimethylaniline (**13**) and haloselenium compound **14** was reacted in glycerol under an inert atmosphere to give organochalcogen molecule **15**. Waste product was HCl and yield of the reaction was excellent, 99%. Furthermore, the energy needed for running of the reaction was minimum, which was room temperature.

When scientists dig an unknown knowledge which has already existed, they have come across unexpected results. One of these situations was reported by Wolfson et al. One of the most important reactions is obviously hydrogenation of organic compounds in which catalysts and molecular hydrogen have been used. Wolfson et al. have achieved hydrogenation of benzaldehyde in glycerol which was used both solvent and hydrogen donor reagent using ruthenium catalyst (**Scheme 5**) [20].

This reaction represents a green protocol as glycerol has been used as both solvent and reagent resulting in atom economic strategy. It was seen that while glycerol oxidized to 1,3-dihydroxy-acetone (**18**), the reaction gave benzyl alcohol (**17**), one of the most important starting materials for organic reactions.

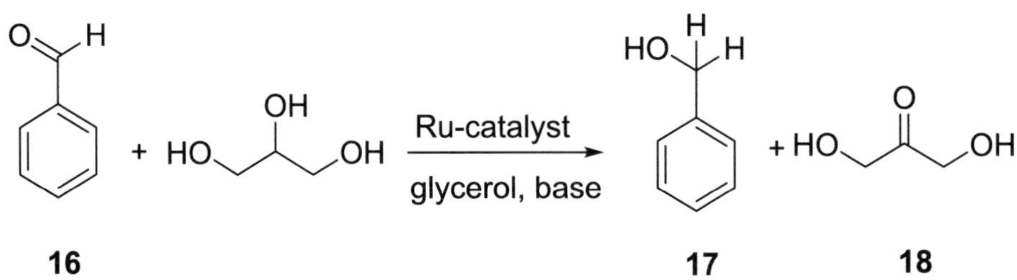

Scheme 5. Hydrogenation of benzaldehyde with catalyst and glycerol [20].

Wolfson et al. have reported the compatibility of glycerol in some named reactions such as nucleophilic substitution, reduction, catalytic reduction, Heck reaction, Asymmetric hydrogenation, and transesterification [21]. They have concluded that glycerol was successfully employed as versatile and alternative green solvent in a variety of organic reactions and synthetic methodologies. In addition, they have said that high products conversions and selectivities were achieved [21].

Functional group protection is still crucial in all fields of industries. Carbonyl protection can be progressed with a reaction between a molecule having a carbonyl group and 1,2- or 1,3-dithiole. Perin et al. have successfully applied a green protocol for protection of carbonyl group. Researchers have done the protection of ketones in glycerol (**Scheme 6**) [22]. Protection of carbonyl group of cyclohexanone (**19**) with 1,2-dithiole (**20**) was achieved in glycerol with good yield, 85% [22].

More recently, Gu et al. have described a cyclization reaction in glycerol in which a three-component strategy was utilized for pyran derivative **25** (**Scheme 7**) to occur [23]. Reactions were run with styrene (**22**), dimedone (**23**), and *p*-formaldehyde (**24**). The yield of the product was based on selected solvent and the most suitable solvent was selected as glycerol in which yield was seen as 68%. The other solvents showed less yield than glycerol. Furthermore, sustainability of glycerol was also tested and after three runs, yield was recorded to be found as 65%. An intermediate product of the reaction between dimedone and p-formaldehyde is formed and this intermediate product forms the pyrane ring by cyclizing with the styrene.

To furnish pyrazolo-pyrane derivative **30**, Lu and his group have designed a reaction in which pyrazolone **28**, styrene analog **29**, and *p*-formaldehyde **24** were reacted in glycerol (**Scheme 8**) [24]. Reaction was progressed at 110°C and yield of the reaction was calculated as 78%. Same reaction in solvent-free and ionic liquids gave no product and was 48%, respectively. Pyrazolone derivative **28** was taken place by the reaction between phenyl hydrazine (**26**) and ethyl acetoacetate (**27**) through well-known condensation. Pyrazolone was not isolated and trapped with styrene analog **29** [24].

Glycerol and its etheric derivatives have been utilized both as a solvent and as a reagent for industrial processes and mg-scale synthetic reactions [25]. Unfortunately, glycerol is not still stood on the top of the industry due to some disadvantages such as viscosity, reactivity, and capability of being a ligand for metals. We hope that chemists and medical scientists will find

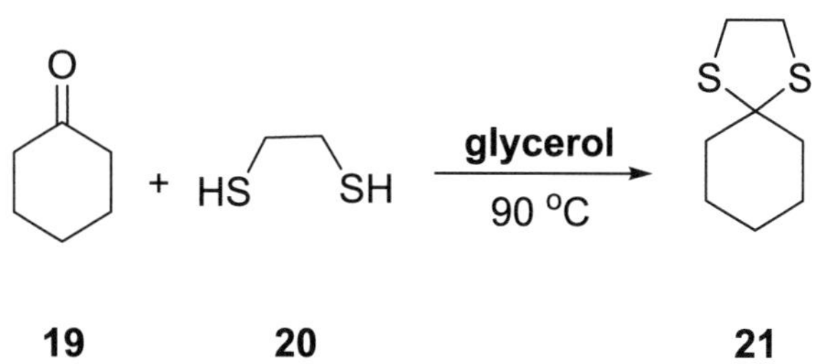

Scheme 6. Protection of carbonyl group with 1,2-dithiole in glycerol [22].

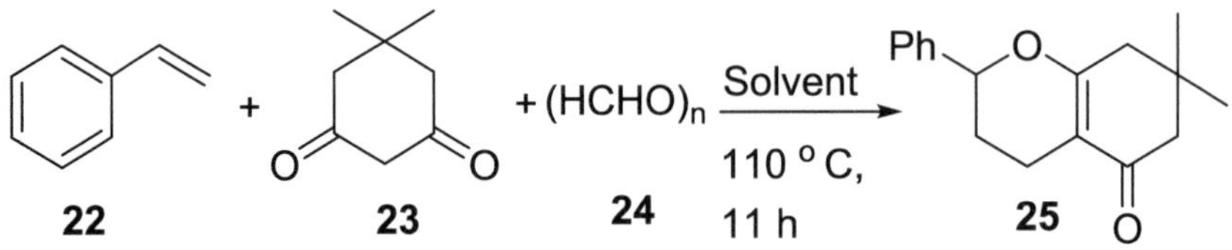

Solvent	Yield
H_2O	14
Toluene	< 10
No solvent	15
Glycerol	68
Reused in the third run	65

25

Scheme 7. Three-component strategy in glycerol in order to get pyrene derivative **25** [23].

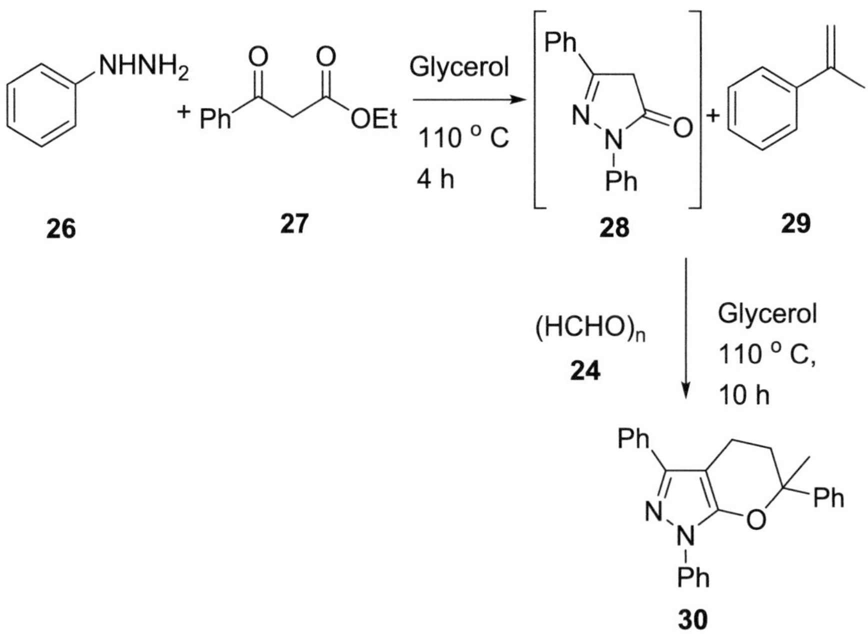

Scheme 8. Synthesis of pyrazolo-pyrane in glycerol [24].

a way for greener alternative. Academicians have an important role in the ability for industry to implement green chemistry while industry can utilize the findings which are reactions, materials, and conditions with industrial relevance, to introduce more sustainable alternatives with lower risk and greener protocols for scale-up productivity.

2.3. Water as a green solvent

Water possesses many unique physical and chemical properties such as extensive hydrogen bonding, high heat capacity, large dielectric constant, and a large temperature window. Water as a solvent has therefore many advantages over conventional organic solvents. Furthermore, water can be selected as a green solvent due to cost, readily available, nontoxic, nonpolluting, and nonflammable. In fact, people do not call water as a chemical. In spite of many important advantages of water, it is still not commonly utilized as a sole solvent for synthetic strategies in research lab and industry as most of the organic compounds are not soluble in water. Nature selects water for its biological reactions and since a century, scientists have tried to mimic the synthetic reactions in water as occurred in nature. Scientists have been away from water for a long time because of an old doctrine in which old chemists say that the insoluble reagents do not yield any product. However, Sharpless has altered this old idea with a new thinking that reactions can be progressed "on" or "in" water means that solubility is not important for reactions [26]. Sharpless has described the reactions such as cycloaddition, Diels-Alder, nucleophilic opening of epoxide and Claisen rearrangement in which as the reactants were not soluble in water, the reactions were described as being on-water [26].

Basic reactions of organic chemistry are utilized in pharmaceutical chemistry to obtain medicines. One of these reactions is Wittig reaction. Wittig reaction is so important and it gives a new C–C bond. Morsch et al. have therefore reported a green protocol for Wittig reaction, run in water at 25°C (**Scheme 9**) [27].

Pyrrole is one of the most critical starting materials for drug design and its reactions are also so important. For a green strategy, Sobral reported a reaction that pyrrole (**34**) and diethyl ketone (**35**) were reacted in water to get 2,2′-dipyrromethane (**36**) (**Scheme 10**) [28]. Sobral has reported that yield of the reaction was 80% and the reaction was progressed as gram-scaled [28].

Synthesis of isocoumarin in H_2O was reported by Xu et al. They have discussed that the reaction of salicylic acid (**37**) and alkyne **38** in the presence of ruthenium catalyst gave isocoumarin with yield of 85% (**Scheme 11**) [29].

Pizzo and coworkers described the reaction of aza compound **40** and vinyl ether **41** resulting in pyridazine derivatives as a sole product with 92% for **42** and 6% for **43** and pyrrole derivative

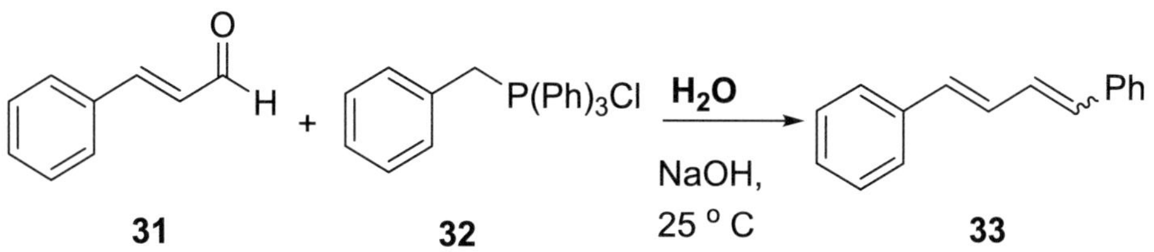

Scheme 9. Wittig reaction in water [27].

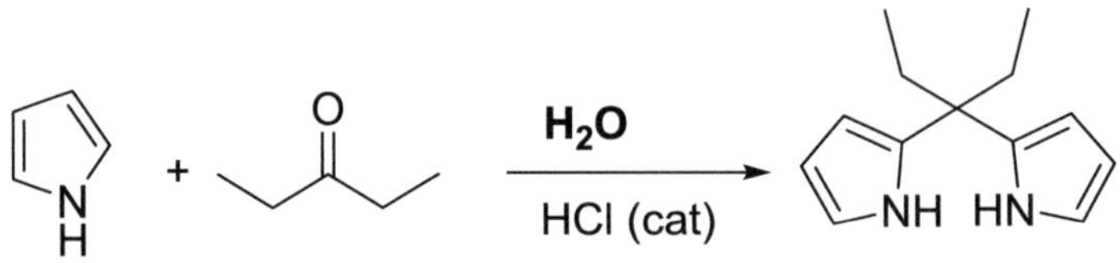

Scheme 10. Dipyrromethane synthesis in water [28].

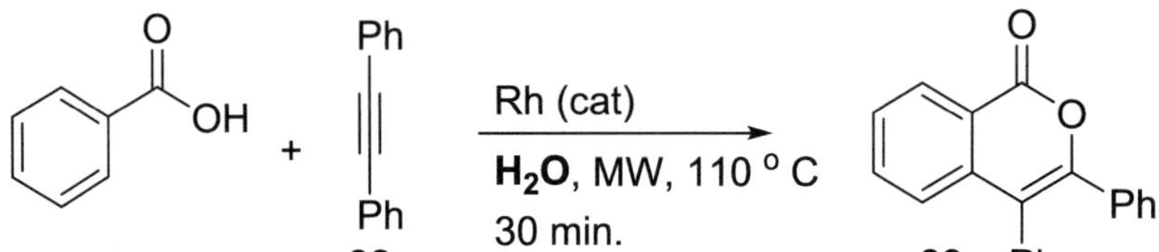

Scheme 11. Synthesis of isocoumarin in water [29].

44 as a by-product with 2%. Authors reported that the reaction proceeded under heterogeneous medium because of poor solubility of aza compound **40** and vinyl ether **41** as called on-water reaction (**Scheme 12**) [30].

Synthesis of benzothiazole ring **47** was published by Patel and co-workers on water [31]. The reaction was started with iodo-benzo-isothiocyanate (**44**) and morpholine (**45**) to get thiourea derivative **46** which was not isolated. Thiourea derivative was cyclized with CuO-nanocatalyst using K_2CO_3 on water with yield of 92% (**Scheme 13**). They have also reported some points which were stereoselectivity, reusable catalyst, and no chromatographic purification because of high yields. They have also screened the effects of different solvents such as dioxane, DMF, and toluene which gave yields of 63, 70, and 55%, respectively [31].

Qu and his group studied the chromene derivative **49** in water. The reaction was named as highly green due to the fact that toxic solvent, catalyst, additive, and base were not used.

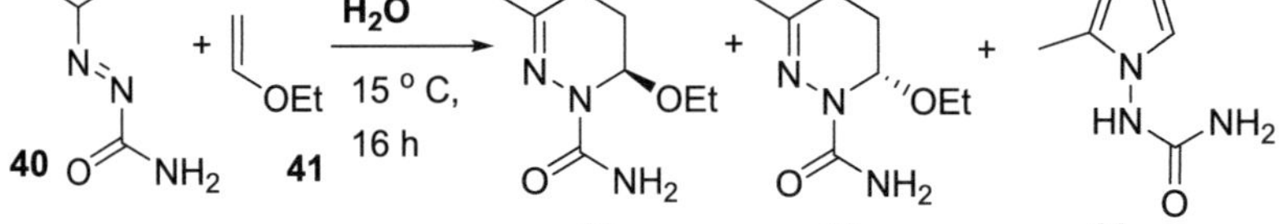

Scheme 12. Diels-Alder reaction of aza compound and vinyl ether "on-water" [30].

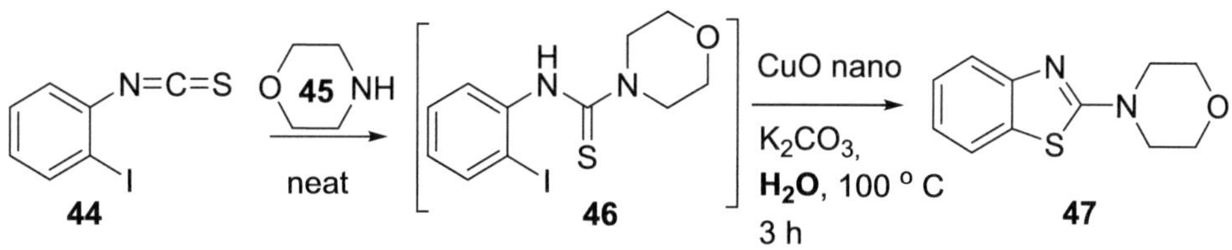

Scheme 13. On water reaction of benzothiazole [31].

Yields of the reactions were varied between 74 and 95% [32]. They have synthesized spiroketal enol derivative **50** with the same strategy using furane ring, as well. Reactions were run in water with the addition of a small amount of hexafluoro-2-propanol (**Scheme 14**).

Lactam is an important ring as it is a part of antibiotic medicines such Cefaclor. With this respect, Pirrung and Sarma have reported the reaction in which acetoacetic acid (**51**), amine **53**, and izonitrile **52** were reacted in water (**Scheme 15**). The reaction was completed in 2 hours with yield of 93%. Authors have described that DCM gave the same result with less yield, 45% [33]. This reaction is a crucial example since it gives lactam in a water medium. Generally speaking, lactam ring is susceptible for ring opening toward nucleophile.

Besides lactam, a lactone is also so crucial ring which scientists want to synthesize. Lactone is presented in many natural and synthetic products which are used as a remedy for diseases. Fujimoto and coworkers studied the cyclization of allyl-iodoacetate using triethylborane in water (**Scheme 16**). Authors claimed that reaction progressed through radical and this is not a recently encountered result [34]. Compatibility of water was tested against some solvents. DMSO, DMF, MeCN, methanol, and benzene gave the cyclic product less than the reaction in water.

Wei and co-workers [35] reported that CH_2 of isatin was converted into oxime group (**Scheme 17**) which is a valuable synthetic building block, present in many bioactive molecules [36].

The reaction was experimented in both water and 1,4-dioxane, and yield of the reaction in water was 96% while the reaction in 1,4-dioxane was 5%. This derivatization was progressed through radical reaction and as said before, the radical reaction in water really challenges and thanks to scientists, these hard reactions have been achieved.

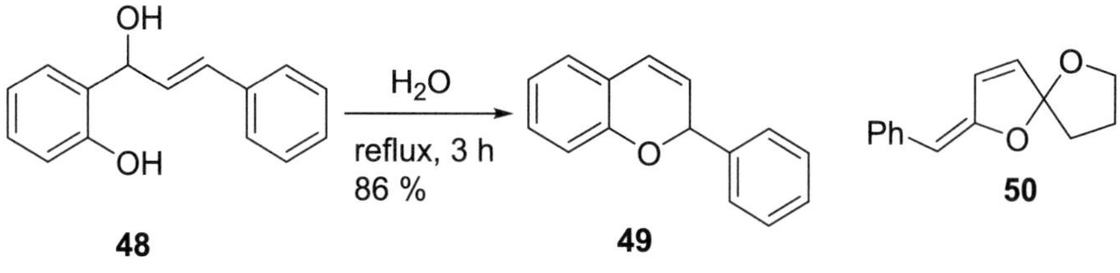

Scheme 14. Synthesize of chromene derivative in water [32].

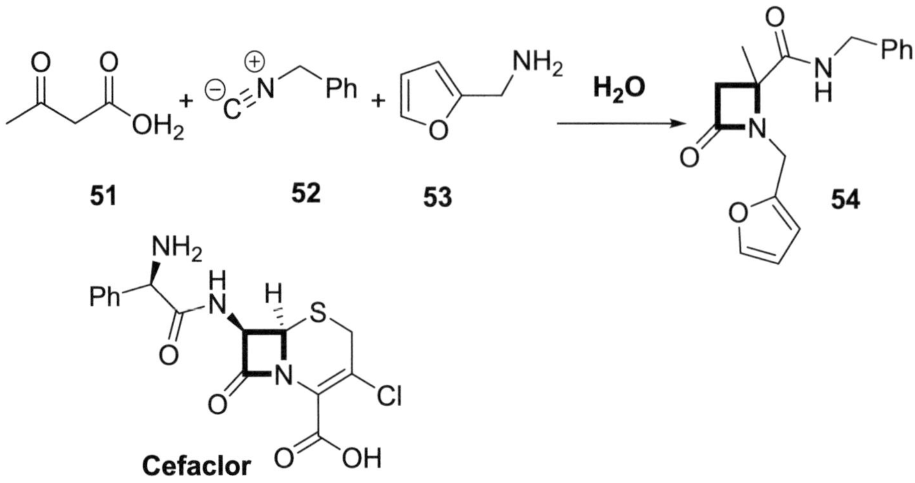

Scheme 15. Synthesis of lactam ring in water [33].

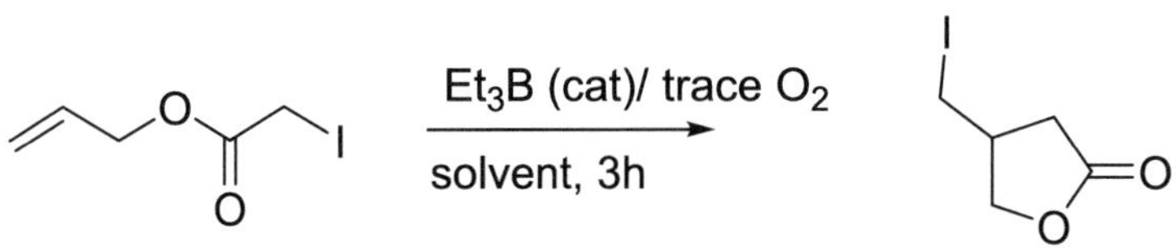

55 56

Solvent	Yield
water	78
DMSO	37
DMF	13
MeCN	13
methanol	6
benzene	<1

Scheme 16. Yielding of lactone derivative [34].

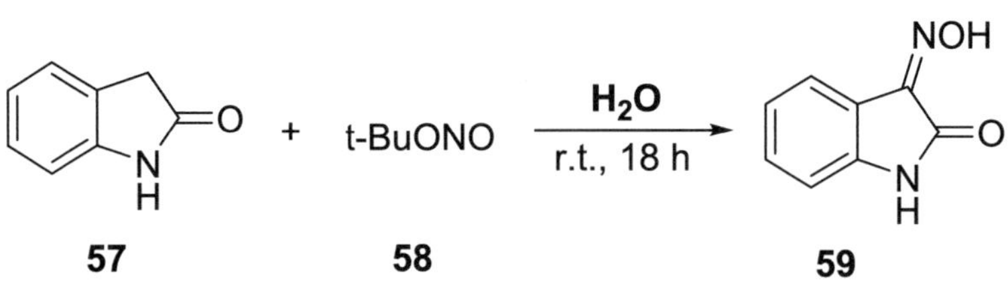

Scheme 17. Conversion of isatin into oxime-isatin **59** [35].

Chen et al. have synthesized C-2 substituted indole derivatives **62** through a reaction in which aniline derivative **60** and organoboron salt **61** were used in water. For the acidity of the medium, tosyl acid was utilized and palladium acetate was consumed as a catalyst. This reaction is one of the promising reactions those run in water. The reaction's yield was calculated as 92% and derivatization of indole was also studied (**Scheme 18**) [37].

Mishra and Verma have introduced a reaction for example of tetracyclic ring. Benzofuran derivative **63** and ortho-phenylendiamine (**64**) were reacted in water to get tetracyclic molecule with 88%. Solvent was tested for yield of the reaction and they have recorded that most of the proper solvents were water and the others were resulted in decreasing of the product down to 45%. Reaction was also studied using $AgNO_3$ and result of catalyzed reaction was unsatisfactory. They have optimized the reaction and explained that there was no need for catalyst, additive, and toxic solvent (**Scheme 19**) [38].

It is obvious that we can not mention all literature about green solvents, but we have desired to focus the most important literature. Exceptions for vegetable oil, glycerol and water have been considered more in detail in which we have given some important synthetic strategies. In next sub-section, we will share some examples of the other solvents which are pointed as green solvents such as ionic liquids, ethyl lactate, and so on.

2.4. The other solvents as mentioned in green chemistry

Ionic liquids are organic salts which are liquid at ambient temperatures. They are nonvolatile, nonflammable, thermally and chemically stable which make them as a better alternative for green chemistry than conventional organic solvents. Due to their high polarity, it can be used in many chemical and biochemical reactions. Besides special properties, they show less solubility in water and are generally immiscible with many organic solvents such hexane(s). They are much more viscous than other organic solvents which might be due to more hydrogen bonds and Van der Waals interactions. The most important feature of ionic liquids is that they can be tuned by changing cation, anion, and alkyl part, in which it is possible to obtain many manipulated green organic solvents [39]. Some common cation and anion parts are presented in **Figure 2**.

On the other hand, to tune the physical properties of ionic liquids, they have combined with hydrogen donor reagents such as glycerol (**68**), oxalic acid (**69**), and urea (**67**). This green alternative emerged because of the volatility of organic solvents. Ionic liquids are called deep eutectic solvents (DES) when composed by a quaternary ammonium salt and a hydrogen bond donor (**Figure 3**) [40]. There are some common combinations for DES which can be seen in **Figure 2**. They have been composed by different amounts of each part.

More recently, ethyl lactate **70** (**Figure 4**) was introduced as a potential green solvent to extract some natural ingredients from vegetable by Gan and co-workers [8].

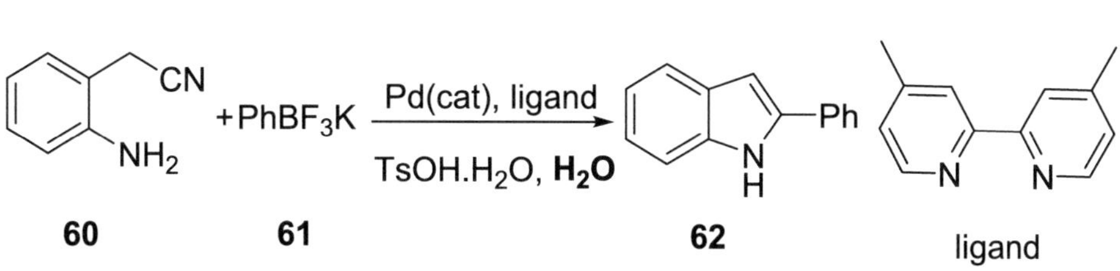

Scheme 18. Synthesis of indole ring [37].

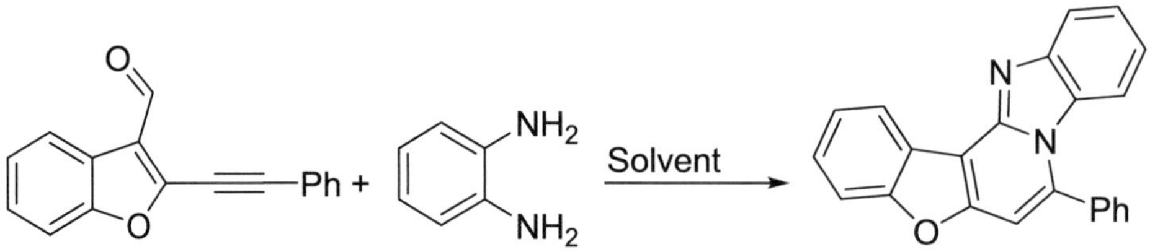

63 64 65

Solvent	Yield
DMSO	45
DMF	50
H_2O	88
Dioxane	40

Scheme 19. Synthesis of tetracyclic molecule **65** in water [38].

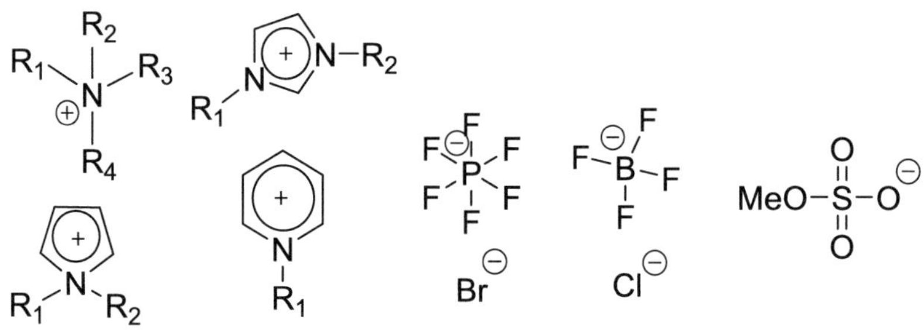

cations for ionic liquids **anions for ionic liquids**

Figure 2. Cations and anions for ionic liquids.

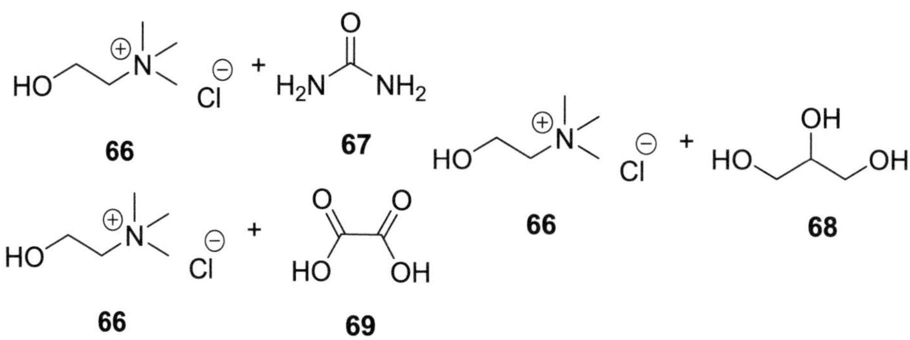

Figure 3. Some of the deep eutectic solvents.

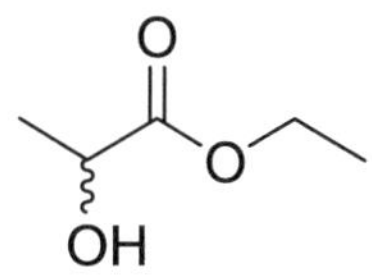

70

Figure 4. The structure of ethyl lactate.

They have studied ethyl lactate for green extraction and they have reported that understanding of its extraction capability and applicability should be improved. This solvent was experimented for plant extraction and might be tested in chemistry and pharmaceutical applications.

3. Synthetic strategies with catalysts

3.1. Nanocatalysts as a green solution

Catalyst is one of the rules of green chemistry and it should be considered by chemists and medical scientists [41]. There are two types of catalysts: heterogeneous and homogeneous. Homogeneous catalysts are more effective to obtain expected products than heterogeneous catalysts. However, isolation and reusable of homogeneous catalysts are the more problematic disadvantages when used for fine chemicals production in the chemical and pharmaceutical industry because of metal contamination of products. Less effective but more attractive heterogeneous catalysts are more favorable due to some of their advantages which are reusable and easier isolation from the medium. Besides heterogeneous catalyst, as a semi-heterogeneous catalyst, nanocatalysts have taken more attention as they have large surface-volume ratio resulting in more interactions between the surface of catalyst and reactant. However, there is still a contamination of catalyst even if it is filtered using specific filtration methods. More recently, thanks to magnetism, magnetic nanocatalysts have been obtained and extracted from medium with external magnetic field [42–46]. They have given more promising solution for chemical industries and it seems to be good candidates for active pharmaceutical ingredient (API) industry [42, 47].

Sharma et al. introduced a cyclization reaction using nanocatalyst [48]. They have obtained oxazole derivatives **73** with the reaction between benzyl amine (**71**) and methyl acetoacetate (**72**). Nanomagnetic catalyst was characterized by SEM, XRD, and FESEM. They have discussed that the yield of oxazole derivative decreased down to 5% absence of the nanocatalyst. Under reaction condition with nanocatalyst, conversion of the reaction was recorded as 100% which means that waste product is not produced (**Scheme 20**).

One-step synthesis of xanthones was achieved by Gerbino and coworkers in which copper-based magnetically recoverable nanocatalyst was utilized [49]. Salicylaldehyde and phenol derivatives were reacted in toluene under ligand-free condition (**Scheme 21**). Reusable copper

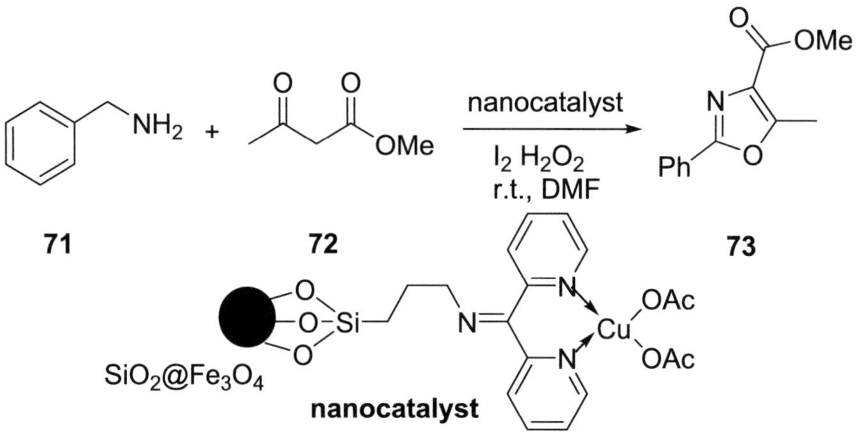

Scheme 20. Synthesize of oxazole derivatives on nanomagnetic catalyst [48].

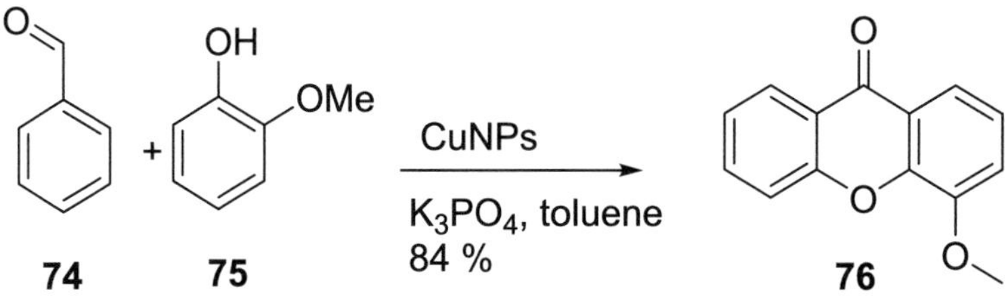

Scheme 21. Furnishing of xanthone derivative **76** with copper-based nanocatalyst [49].

nanocatalyst was tested and was found to be 89% effective when used in fourth cycle. Altering of copper nanocatalyst with a conventional catalyst, CuCl or CuO, decreased the yield of the product down to 65 and 62%, respectively.

Quinazolinones **79** were formed with halo benzamide **77** and benzylamine using copper nanocatalyst (CuONPs) by Patel et al. (**Scheme 22**) [50]. Researchers showed that conventional copper catalysts such as CuBr, CuCl, and CuI have a less catalytic effect than copper nanocatalysts. Furthermore, without a catalyst, there is no cyclic product.

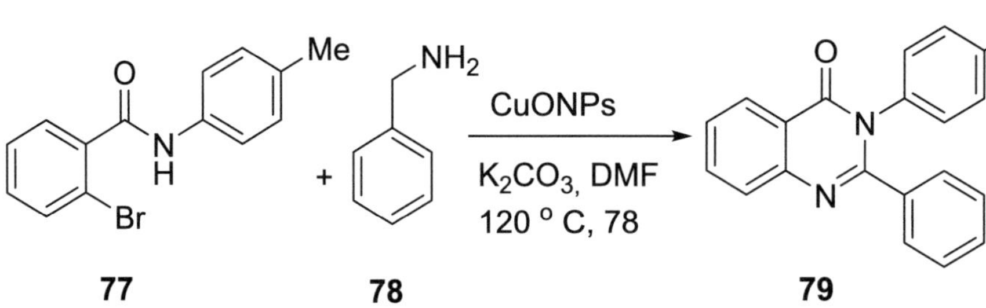

Scheme 22. Synthesis of quinazolinone ring [50].

Some of the most important synthetic strategies are named Suzuki, Heck, and Sonogashira reactions. These strategies give countless methods to chemists and medical scientists for further reactions in which fine chemicals and medicines can be obtained easily and in greener ways. More recently, these reactions have been progressed with magnetic nanocatalyst which are mentioned green strategies [51–56].

A green alternative for hydrogenation of π bonds and the nitro group was reported by Baig and Varma [57]. They have experimented lots of double and triple bonds, and nitro groups for reduction to obtain saturated alkyl unit and amine functional groups (**Scheme 23**). For fine chemicals, reduction is a critical reaction and thanks to this reaction, reduction can be done by a nanocatalyst which is magnetically active and can be removed easily with an external magnetic field from reaction medium [57].

N-aryl oxazolidine-2-ones framework **82** is an important ring which is a part of some clinically used medicines such as linezolid and Rivaroxaban which are sold as antibiotic and anticoagulant, respectively. In this spirit, forming of these rings with green chemistry would be very useful. Sharma's group has therefore studied on that strategy to obtain N-aryl-oxazolidine-2-ones using magnetic nanocatalyst (**Scheme 24**) [58]. They have tested nanocatalyst for reusability and seen that after eight runs, the yield of the product was recorded as 80–85% which was close to the first run, 95% [58]. Anilin (**80**) and ethylene carbonate (**81**) were reacted by means of nanocatalyst and the reaction gave almost quantitative yield. Removing of the nanocatalyst is so easy due to its magnetic feature. The nanocatalyst was made with iron (III) and iron (II) salts and the nano-iron oxide was reacted with silicone derivatives to obtain a silicon-coated nano-iron oxide. After then, imidazole-terminated silicon derivative was bonded to nano-iron oxide. Authors have assumed that nanocatalyst bonds to the oxygen atom of the carbonyl group of carbonate and makes therefore easy attack of aniline to a carbon atom of carbonate.

Benzodiazepine substructure is presented in some vital medicines such as diazepam, alprazolam, lorazepam, oxazepam, temazepam, and clonazepam. This structure can be obtained by many different methods [59, 60]. Beside presented methods, green chemistry is in progress to give diazepine derivatives. Lutfullah et al. have published an article in which they have displayed a green reaction using nanocatalyst to obtain tricyclic benzodiazepine derivatives **85** in good yields (**Scheme 25**). Nanocatalyst is silicon-coated nickel-oxide [61].

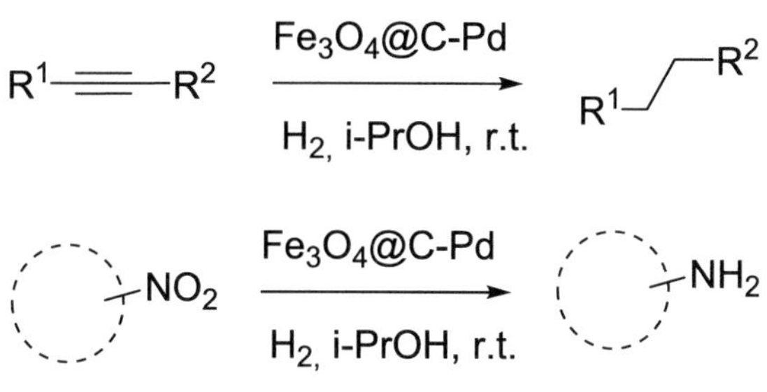

Scheme 23. Reduction reactions with magnetic nanocatalyst [57].

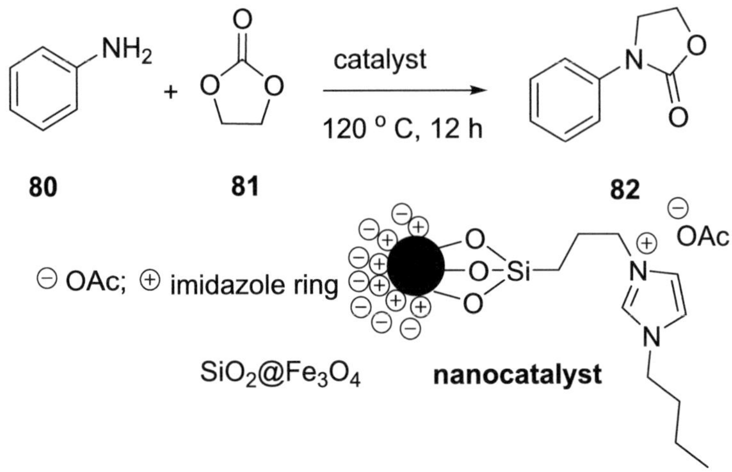

Scheme 24. N-aryl-oxazolidin-2-one with nanocatalyst [58].

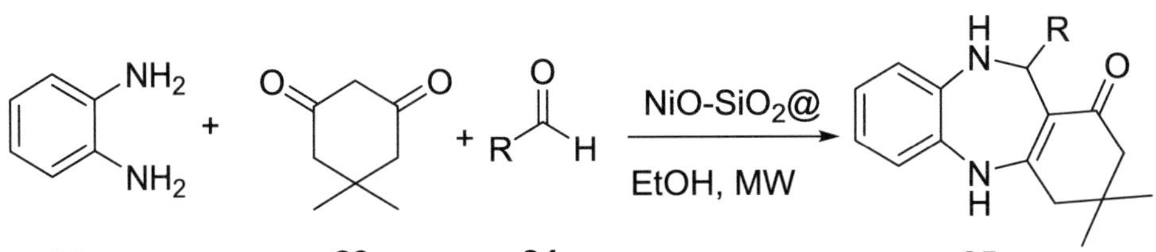

Scheme 25. Synthesis of diazepine derivatives [61].

1,2-diamino benzene (**83**), dimedone (**23**), and aromatic aldehydes (**84**) were reacted in the reaction tube in which nanocatalyst was presented. All reactions were run in microwave synthesizer which is a good tool for green chemistry.

3.2. Biocatalysts as a green solution

In the literature, biocatalyst is a biological material which can be an isolated enzyme, a crude cell-free extract, an immobilized enzyme, or enzymes in whole microbial cells. Enzymes are critical endogens and have a vital role in living cells, which catalyze all the in vivo metabolic reactions to produce a necessary product for the body. To mimic the activity of enzymes for our reactions, since a century, enzymes have been utilized for the reactions in the laboratory. For many different purposes, scientists have used enzymes which are oxidoreductases, transferases, hydrolases, lyases, isomerases, and ligases. These enzymes are utilized in the industry such as food, pharmacology, medicine, and textiles. Enzymes have unique properties which can sometimes not be mimicked by artificial organic products. Enzymes show highly stereoselectivity resulting in purely one isomer and can therefore decrease the cost of medicine because chirality has a high effect on the medicine cost when candidate medicine has more than one chiral center. This chemical potential forces the chemists and medical scientists to design biocatalysts to put them into the reaction flask [62, 63].

Biocatalysts are biodegradable, sustainable, reusable, more efficient, and more stereoselective which means more atom economic than conventional methods. With this respect, they are more powerful tools for green chemistry.

As mentioned before, chirality is one of the most critical criteria for pharmaceutical cost and pharmacologic effect of medicine which has a chiral center(s).It is well known that thalidomide tragedy revealed due to a racemic mixture of drug. Enzymes hence show up to solve this problem. Complete conversion of racemic amino acid amides to optically active amino acid derivatives was studied using lipase/Pd catalyst via dynamic kinetic resolution [64]. Authors have described that the reaction provided good yields (80–98%) and high enantiomeric excess (95–98% ee) (**Scheme 26**) [64].

Savile and co-workers have reported an efficient biocatalytic process to replace a recently implemented Rh-catalyzed asymmetric enamine hydrogenation for antidiabetic compound sitagliptin. Current synthesis of sitagliptin involves enamine formation followed by asymmetric hydrogenation at high pressure using Rh-based chiral catalyst in which sitagliptin was formed in 97% ee with trace amount of Rh [65]. Savile's synthetic route showed green reaction that is direct amination of prositagliptin ketone **88** to furnish enantiopure sitagliptin **89** (99.95% ee) followed by phosphate salt formation to get sitagliptin phosphate **90** (**Scheme 27**) [65].

Ghiladi et al. have emerged a biocatalytic green process for the oxidation of pyrrole ring to provide pyrroline-2-one (**92**) [66]. Dehaloperoxidase (DHP) was supplied for biocatalytic reaction and for oxygen source, hydrogen peroxide was consumed. Finally, pyrroline-2-one was obtained with 31.7% conversion (**Scheme 28**). Authors explained that some derivatives of pyrrole were oxidized up to 100% conversion.

Chen and co-workers experimented to improve the performance of immobilized lipase by interfacial activation on iron-oxide nanoparticles. They have tested immobilized enzyme stability and displayed that immobilized lipase exhibited much better stabilities [67]. Furthermore, with iron-oxide nanoparticles, the enzyme was removed easily as mentioned in magnetic nanocatalyst.

Besides iron oxide nanoparticles, enzyme immobilization on carbon nanotubes (CNTs) and graphene is applied for many chemical reactions such as cyclization, selective amination, trans esterification, and redox reactions and biosensing applications for detection of glucose, phenol, and hydrogen peroxide [62, 63].

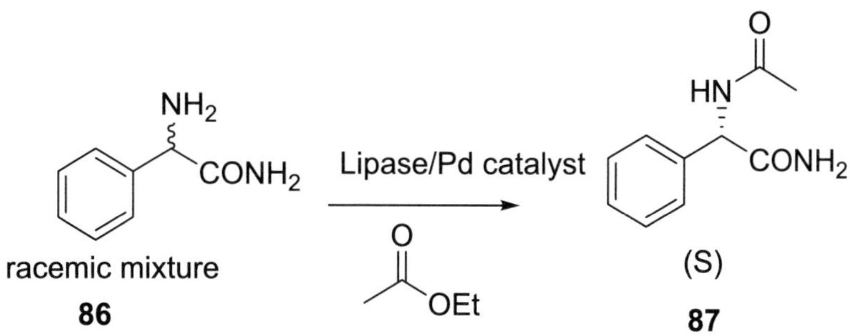

Scheme 26. Resolution of racemic amino acid amides by lipase/Pd [64].

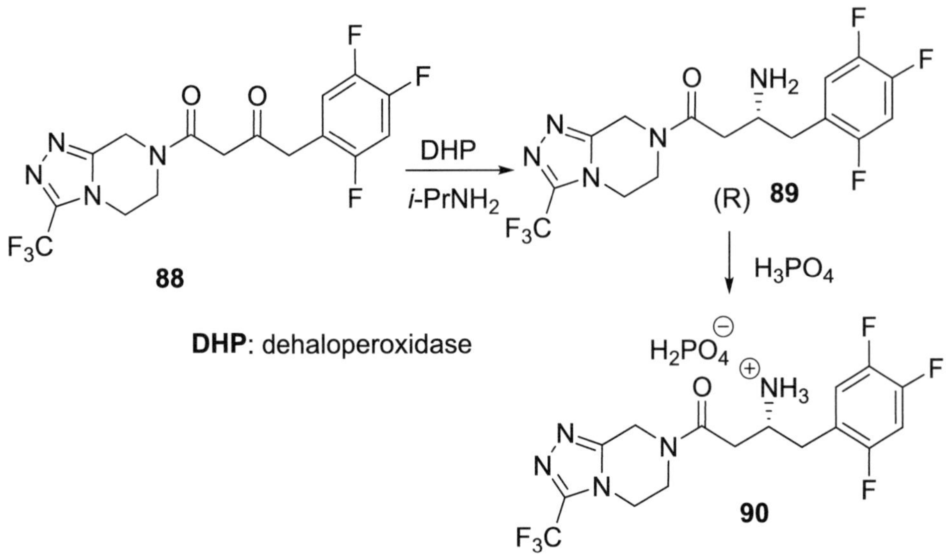

Scheme 27. Synthesis of sitagliptin phosphate with green chemistry.

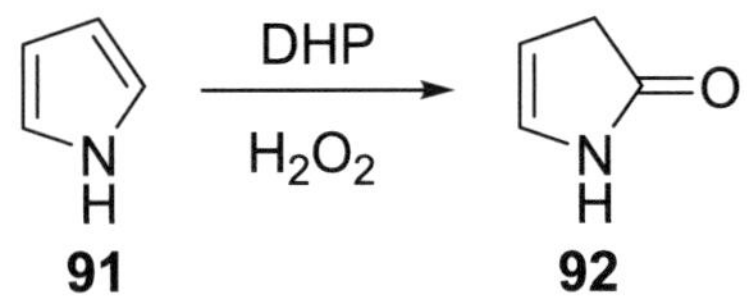

Scheme 28. Oxidation of pyrrole by DHP.

Kroutil et al. reported the formation of lactam ring **94** starting from 4-oxo ester **93** in which lactate dehydrogenase, transaminase, and D-alanine were added to the reaction medium. The reaction was progressed at 30°C and 7.0 pH. Reaction was finalized to obtaining lactam derivative with 92% yield (**Scheme 29**) [68].

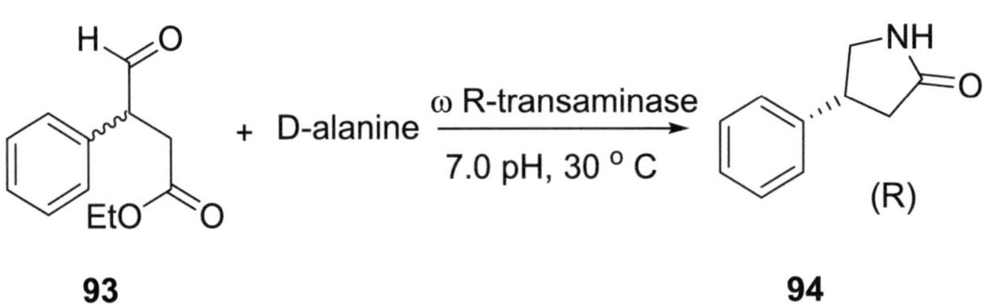

Scheme 29. Cyclization of keto-ester to lactam by enzyme.

4. Case studies in green chemistry

In 2002, Pfizer won the U.S. Presidential Green Chemistry Award for alternative synthetic pathways for its innovative manufacturing process for sertraline hydrochloride which is the active ingredient of Zoloft that is used to treat clinical depression. Furthermore, pregabalin, sold as Lyrica, is manufactured by Pfizer for the management of neuropathic pain and epilepsy. Pfizer has designed a green route for the synthesis of pregabalin and they have reported that almost 38 million liters of alcoholic organic solvents and nearly 2000 metrics tons of raw materials were eliminated on an annual basis. Their achievements were based on biocatalyst, lipase, which was used for resolution of cyano diester **95** (**Scheme 30**) [69].

GlaxoSmithKline company has announced a green reaction way for paroxetine, sold Seroxat, and Paxil, which is used for anxiety disorder. They have discussed that the yield of the overall transformation was almost double that of the process in conventional route, resulting in a greener, shorter, and more cost-efficient way. Critical step was applying of protease enzyme which was regioselectively hydrolyzed an ester group (**Scheme 31**) [69].

Sertraline hydrochloride as known Zoloft is a selective inhibitor of serotonin reuptake which is utilized for the curing of depression [70]. When it was synthesized by conventional synthetic route, chemical reagents and metal salts were consumed. On the other hand, removing of metal salts and Pd/C catalyst gave more selective and greener protocol (**Scheme 32**) [71].

Colberg and co-workers in Pfizer Global Research and Development have designed a green protocol for sertraline in which toxic solvents such as toluene, hexane were removed from

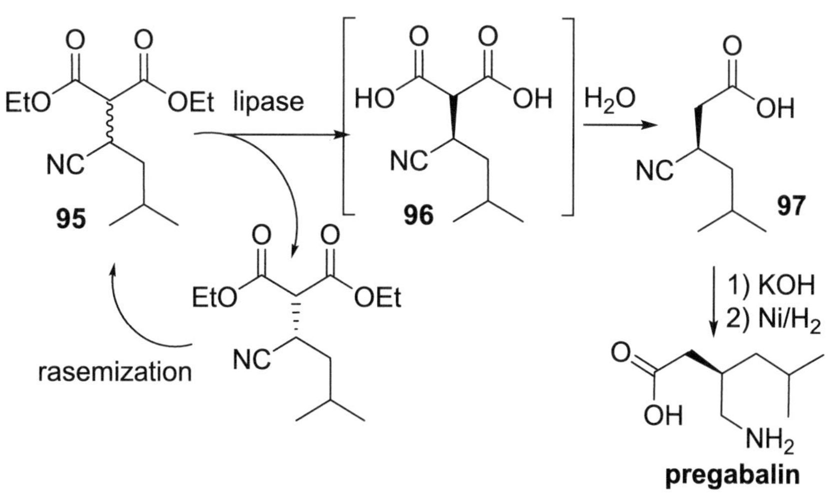

Scheme 30. Green protocol for pregabalin.

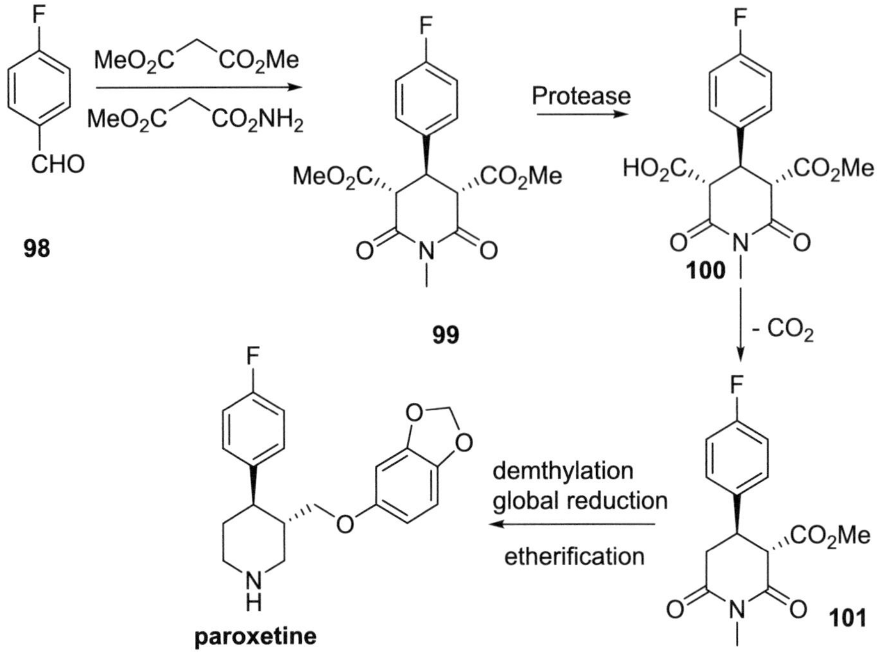

Scheme 31. Formation of paroxetine through biocatalytic route.

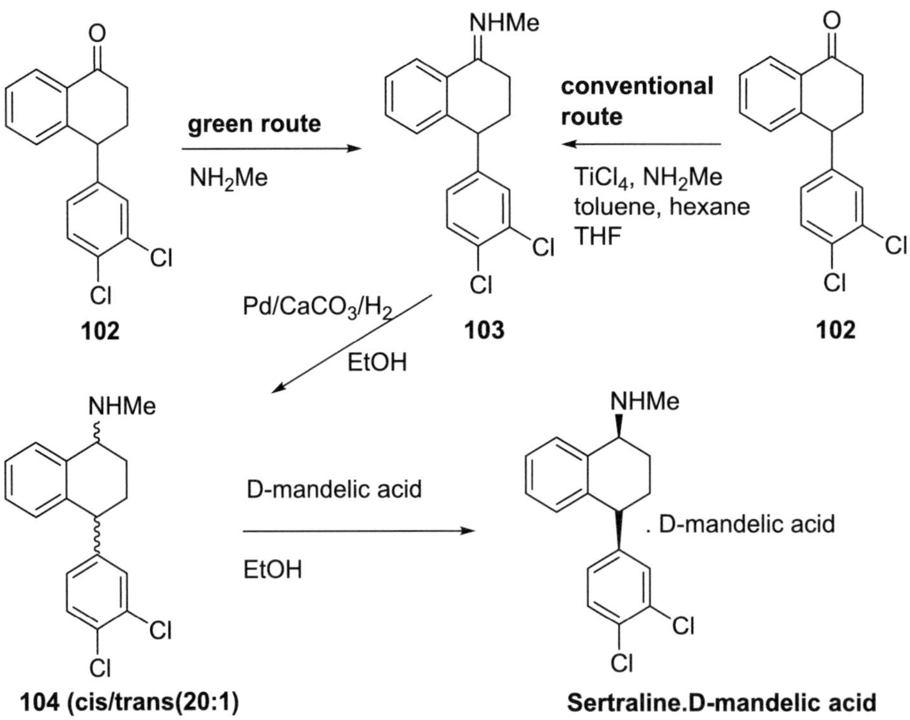

Scheme 32. Green protocol for sertraline [71].

the strategy and comparison of solvent utilization between the first commercial route and the new green route showed that 76.000 L solvents, 440 tons/year of TiO_2-$MeNH_2$.HCl waste and about 40 tons of the unwanted trans isomer waste were eliminated [71].

5. Conclusion

Green chemistry is getting extended in many researches and industry areas. The reason is that the resources of the world are limited and it is necessary to be consumed with caution. On the other hand, we have already witnessed that researchers and pharmaceutical companies searched out for green protocol when manufactured the pharmaceuticals. In this spirit, most pharmaceutical companies are making increasing efforts to limit waste and avoid air and water pollution. Green solvents, nanocatalysts, and biocatalysts give many opportunities for greener methods in which impact on the environment and the cost of pharmaceuticals can be decreased. We hope that this chapter and the others give a brief consideration of importance of green chemistry. With advantages of green chemistry, hopefully, industry will alter conventional methods with greener ones.

Author details

Nurettin Menges

Address all correspondence to: nurettinmenges@yyu.edu.tr

Faculty of Pharmacy, Van Yuzuncu Yil University, Van, Turkey

References

[1] Green Chemical Industry to Soar to USD 98.5 Billion by 2020, Navigant Research, Jun 20, 2011

[2] Nature Comment. Nature. Jun 2, 2016;**534**:27-29

[3] Stephen KR. C&EN Washington C&EN. Jul 4, 2016;**94**(27):22-25

[4] Sheldon RA. Catalysis: The key to waste minimization. Journal of Chemical Technology & Biotechnology. 1997;**68**:381-388

[5] Prat D, Pardigon O, Flemming H-W, Letestu S, Ducandas V, Isnard P, Guntrum E, Senac T, Ruisseau S, Cruciani P, Hosek P. Sanofi's solvent selection guide: A step toward more sustainable processes. Organic Process Research & Development. 2013;**17**:1517-1525

[6] Prat D, Hayler J, Wells A. A survey of solvent selection guides. Green Chemistry. 2014;**16**:4546

[7] Henderson RK, Jimenez-Gonzalez C, Constable DJC, Alston SR, Inglis GGA, Filsher G, Sherwood J, Binks SP, Curzons AD. Expanding GSK's solvent selection guide – Embedding sustainability into solvent selection starting at medicinal chemistry. Green Chemistry. 2011;**13**:854-862

[8] Kua YL, Gan S, Morris A, Kiat H. Ethyll actate as a potential green solvent to extract hydrophilic (polar) and lipophilic (non-polar) phyto nutrients simultaneously from fruit and vegetable by-products. Sustainable Chemistry and Pharmacy. 2016;**4**:21-31

[9] Simon M-O, Li C-J. Green chemistry oriented organic synthesis in water. Green Chemistry. 2012;**41**:1415-1427

[10] Capello C, Fischer U, Hungerbühler K. What is a green solvent? A comprehensive framework for the environmental assessment of solvents. Green Chemistry. 2007;**9**:927-934

[11] O'Brien RD. Fats and Oils: Formulating and Processing for Applications. 3rd ed. CRC Press; Boca Raton, USA; 2005

[12] Menges N, Şahin E. Metal- and base-free combinatorial reaction for C-acylation of 1,3-diketo compounds in vegetable oil: The effect of natural oil. ACS Sustainable Chemistry & Engineering. 2014;**2**:226-230

[13] Claude S. Research of new outlets for glycerol – recent developments in France. Fett/ Lipid. 1999;**101**:101-104

[14] Safaei HR, Shekouhy M, Rahmanpur S, Shirinfeshan A. Glycerol as a biodegradable and reusable promoting medium for the catalyst-free one-pot three component synthesis of 4H-pyrans. Green Chemistry. 2012;**14**:1696-1704

[15] Gu Y, Barrault J, Jerome F. Glycerol as an efficient promoting medium for organic reactions. Advanced Synthesis and Catalysis. 2008;**350**:2007-2012

[16] Garcia JI, Garcia-Marin H, Mayoral JA, Perez P. Green solvents from glycerol. Synthesis and physico-chemical properties of alkyl glycerol ethers. Green Chemistry. 2010;**12**:426-434

[17] Trippier PC, LAbby KJ, Hawker DD, Mataa JJ, Silverman RB. Target- and mechanism-based therapeutics for neurodegenerative diseases: Strength in numbers. Journal of Medicinal Chemistry. 2013;**56**:3121-3147

[18] Moussaoui S, Obinu M-C, Daniel N, Reibaud M, Blanchard V, Imperato A. The antioxidant ebselen prevents neurotoxicity and clinical symptoms in a primate model of Parkinson's disease. Experimental Neurology. 2000;**166**:235-245

[19] Wang Z, Li W, Wang Y, Li X, Huang L, Li X. Design, synthesis and evaluation of clioquinol–ebselen hybrids as multi-target-directed ligands against Alzheimer's disease. RSC Advances. 2016;**6**:7139

[20] Tavor D, Sheviev O, Dlugy C, Wolfson A. Transfer hydrogenations of benzaldehyde using glycerol as solvent and hydrogen source. Canadian Journal of Chemistry. 2010;**88**:305-308

[21] Wolfson A, Dlugy C, Shotland Y. Glycerol as a green solvent for high product yields and selectivities. Environmental Chemistry Letters. 2007;**5**:67-71

[22] Perin G, Mello LG, Radatz CS, Savegnago L, Alves D, Jacob RG, Lenardao EJ. Green, catalyst-free thioacetalization of carbonyl compounds using glycerol as recyclable solvent. Tetrahedron Letters. 2010;**51**:4354-4356

[23] Li M, Chen C, He F, Gu Y. Multicomponent reactions of 1,3-cyclohexanediones and formaldehyde in glycerol: Stabilization of paraformaldehyde in glycerol resulted from using dimedone as substrate. Advanced Synthesis and Catalysis. 2010;**352**:519-530

[24] Tan J-N, Li M, Gu Y. Multicomponent reactions of 1,3-disubstituted 5-pyrazolones and formaldehyde in environmentally benign solvent systems and their variations with more fundamental substrates. Green Chemistry. 2010;**12**:908-914

[25] Pagliaro M, Rossi M. The future of glycerol. In: RSC Green Chemistry Series. 2nd ed. RSC Publishing, Cambrige; 2010

[26] Narayan S, Muldoon J, Finn MG, Fokin VV, Kolb HC, Sharpless KB. "On Water": Unique reactivity of organic compounds in aqueous suspension. Angewandte Chemie International Edition. 2005;**44**:3275-3279

[27] Morsch LA, Deak L, Tiburzi D, Schuster H, Meyer B. Green aqueous Wittig reaction: Teaching green chemistry in organic teaching laboratories. Journal of Chemical Education. 2014;**91**:611-614

[28] Sobral AJFN. Synthesis of meso-diethyl-2,2′-dipyrromethane in water. Journal of Chemical Education. 2006;**83**:1665-1666

[29] Li Q, Yan Y, Wang X, Gong B, Tang X, Shi J, Xu HE, Yi W. Water as a green solvent for efficient synthesis of isocoumarins through microwave-accelerated and Rh/Cu-catalyzed C–H/O–H bond functionalization. RSC Advances. 2013;**3**:3402

[30] Attanasi OA, Crescentini L, Filippone P, Fringuelli F, Mantellini F, Matteucci M, Piermatti O, Pizzo F. Inverse-electron-demand Diels-Alder reactions of (E)-3-diazenylbut-2-enes in water. Helvetica Chimica Acta. 2001;**84**:513

[31] Kumar S, Guin S, Nath J, Patel BK. An "on-water" exploration of CuO nanoparticle catalysed synthesis of 2-aminobenzothiazoles. Green Chemistry. 2012;**14**:2491

[32] Zhang F-Z, Tian Y, Li G-X, Qu J. Intramolecular etherification and polyene cyclization of π-activated alcohols promoted by hot water. The Journal of Organic Chemistry. 2015;**80**: 1107-1115

[33] Pirrung MC, Sarma KD. Multicomponent reactions are accelerated in water. Journal of the American Chemical Society. 2004;**126**:444-445

[34] Yorimitsu H, Nakamura T, Shinokubo H, Oshima K, Omoto K, Fujimoto H. Powerful solvent effect of water in radical reaction: Triethylborane-induced atom-transfer radical cyclization in water. Journal of the American Chemical Society. 2000;**122**:11041-11047

[35] Wei W-T, Zhu W-M, Ying W-W, Huang Y-L, Liang H. Metal-free synthesis of isatin oximes via radical coupling reactions of oxindoles with t-BuONO in water. Organic & Biomolecular Chemistry. 2017;**15**:5254-5257

[36] Pinto AC, Lapis AAM, Silva BV, Bastos RS, Dupont J, Neto BAD. Pronounced ionic liquid effect in the synthesis of biologically active isatin-3-oxime derivatives under acid catalysis. Tetrahedron Letters. 2008;**49**:5639

[37] Yu S, Hu K, Gong J, Qi L, Zhu J, Zhang Y, Cheng T, Chen J. Palladium-catalyzed tandem addition/cyclization in aqueous medium: Synthesis of 2-arylindoles. Organic & Biomolecular Chemistry. 2017;**15**:4300-4307

[38] Mishra PK, Verma AK. Metal-free regioselective tandem synthesis of diversely substituted benzimidazo-fused polyheterocycles in aqueous medium. Green Chemistry. 2016;**18**:6367-6372

[39] Yang Z, Pan W. Ionic liquids: Green solvents for nonaqueous biocatalysis. Enzyme and Microbial Technology. 2005;**37**:19-28

[40] Lopes EF, Gonçalves LC, Vinueza JCG, Jacob RG, Perin G, Santi C, Lenardo EJ. DES as a green solvent to prepare 1,2-bis-organylseleno alkenes. Scope and limitations. Tetrahedron Letters. 2015;**56**:6890-6895

[41] Sun Q, Zhang X-Q, Wang Y, Lu A-H. Recent progress on core-shell nanocatalysts. Chinese Journal of Catalysis. 2015;**36**:683-691

[42] Wang D, Astruc D. Fast-growing field of magnetically recyclable nanocatalysts. Chemical Reviews. 2014;**114**:6949-6985

[43] Polshettiwar V, Luque R, Fihri A, Zhu H, Bouhrara M, Basset J-M. Magnetically recoverable nanocatalysts. Chemical Reviews. 2011;**111**:3036-3075

[44] Hu M, Yao Z, Wang X. Graphene-based nanomaterials for catalysis. Industrial & Engineering Chemistry Research. 2017;**56**:3477-3502

[45] Liu J, Sun W. Magnetic nanoparticles in catalysis, magnetic nanomaterials: Applications in catalysis and life sciences, Chapter 3. RSC Publishing, Cambrige. 2017:59-98

[46] Pena L, Wang D. Recyclable magnetic materials for biomass conversion, magnetic nanomaterials: Applications in catalysis and life sciences, Chapter 5. The Royal Society of Chemitry. 2017:120-141

[47] Ciriminna R, Pagliaro M. Green chemistry in the fine chemicals and pharmaceutical industries. Organic Process Research and Development. 2013;**17**:1479-1484

[48] Dutta S, Sharma S, Sharma A, Sharma RK. Fabrication of core–shell-structured organic–inorganic hybrid nanocatalyst for the expedient synthesis of polysubstituted oxazoles via tandem oxidative cyclization pathway. ACS Omega. 2017;**2**:2778-2791

[49] Menendez CA, Nador F, Radivoy G, Gerbino DC. One-step synthesis of xanthones catalyzed by a highly efficient copper-based magnetically recoverable nanocatalyst. Organic Letters. 2014;**16**:2846-2849

[50] Modi A, Ali W, Mohanta PR, Khatun N, Patel BK. CuO nanoparticle catalyzed synthesis of 2,3-disubstituted quinazolinones via sequential N-arylation and oxidative C–H amidation. ACS Sustainable Chemistry & Engineering. 2015;**3**:2582-2590

[51] Baruwati B, Guin D, Manorama SV. Pd on surface-modified $NiFe_2O_4$ nanoparticles: A magnetically recoverable catalyst for Suzuki and Heck reactions. Organic Letters. 2007;**9**:5377-5380

[52] Yinghuai Z, Peng SC, Emi A, Zhenshun S, Monalisa M, Kemp RA. Supported ultra small palladium on magnetic nanoparticles used as catalysts for Suzuki cross-coupling and Heck reactions. Advanced Synthesis and Catalysis. 2007;**349**:1917-1922

[53] Ko S, Jang J. A highly efficient palladium nanocatalyst anchored on a magnetically functionalized polymer-nanotube support. Angewandte Chemie International Edition. 2006;**45**:7564-7567

[54] Zheng Y, Stevens PD, Gao Y. Magnetic nanoparticles as an orthogonal support of polymer resins: Applications to solid-phase Suzuki cross-coupling reactions. The Journal of Organic Chemistry. 2005;**71**:537-542

[55] Stevens PD, Li G, Yen M, Gao Y. Recycling of homogeneous Pd catalysts using superparamagnetic nanoparticles as novel soluble supports for Suzuki, Heck, and Sonogashira cross-coupling reactions. Chemical Communications. 2005;**35**:4435-4437

[56] So M-H, Liu Y, Ho C-M, Che C-M. Graphite-supported gold nanoparticles as efficient catalyst for aerobic oxidation of benzylic amines to imines and N-substituted 1,2,3,4-tetrahydroisoquinolines to amides: Synthetic applications and mechanistic study. Chemistry, an Asian Journal. 2009;**4**:1551-1561

[57] Baig RBN, Varma RS. Magnetic carbon-supported palladium nanoparticles: An efficient and sustainable catalyst for hydrogenation reactions. ACS Sustainable Chemistry & Engineering. 2014;**2**:2155-2158

[58] Gupta R, Yadav M, Gaur R, Arora G, Sharma RK. Straightforward one-pot synthesis of bioactive N-aryl oxazolidin-2-ones via highly efficient $Fe_3O_4@SiO_2$-supported acetate-based butylimidazolium ionic liquid nanocatalyst under metal- and solvent-free conditions. Green Chemistry. 2017;**19**:3801-3812

[59] Zhang A, Neumeyer JL, Baldessarini RJ. Recent progress in development of dopamine receptor subtype-selective agents: Potential therapeutics for neurological and psychiatric disorders. Chemical Reviews. 2007;**107**:274-302

[60] Antonow D, Thurston DE. Synthesis of DNA-interactive pyrrolo[2,1-c][1,4]benzodiazepines (PBDs). Chemical Reviews. 2011;**111**:2815-2864

[61] Nasir Z, Ali A, Shakir M, Wahab R, Lutfullah S. Silica supported NiO nanocomposite prepared via sol-gel technique and its excellent catalytic performance for one-pot multicomponent synthesis of benzodiazepine derivatives under microwave irradiation. New Journal of Chemistry. 2017;**41**:5893-5903

[62] Botta L, Bizzarri BM, Crucianelli M, Saladino R. Advances in biotechnological synthetic applications of carbon nanostructured systems. Journal of Materials Chemistry B. 2017;**5**:6490-6510

[63] Liu W-L, Lirio S, Huang H-Y, Ma S. Functional metal-organic frameworks for enzyme/protein immobilization, functional supramolecular materials: From surfaces to MOFs, Chapter 8. RSC Publishing, Cambrige. 2017:281-296

[64] Choi YK, Kim Y, Park J, Kim M-J. Synthesis of optically active amino acid derivativesvia dynamic kinetic resolution. The Journal of Organic Chemistry. 2009;**74**:9543-9545

[65] Savile CK, Janey JM, Mundorff EC, Moore JC, Tam S, Jarvis WR, Colbeck JC, Krebber A, Fleitz FJ, Brands J, Devine PN, Huisman GW, Hughes GJ. Biocatalytic asymmetric synthesis of chiral amines from ketones applied to sitagliptin manufacture. Science. 2010;**329**:305-309

[66] McCombs NL, Smirnova T, Ghiladi RA. Oxidation of pyrrole by dehaloperoxidase-hemoglobin: Chemoenzymatic synthesis of pyrrolin-2-ones. Catalysis Science & Technology. 2017;**7**:3104-3118

[67] Chen Z, Liu L, Yang R. Improved performance of immobilized lipase byinterfacial activation on Fe_3O_4@PVBC nanoparticles. RSC Advances. 2017;**7**:35169

[68] Koszelewski D, Clay D, Faber K, Kroutil W. Synthesis of 4-phenylpyrrolidin-2-one via dynamic kinetic resolution catalyzed by transaminases. Journal of Molecular Catalysis B: Enzymatic. 2009;**60**:191-194

[69] Koenig SG. Scalable Green Chemistry, Case Studies fron the Pharmaceutical Industry. Taylor & Francis Group, CRC Press; Boca Raton, USA. 2013

[70] MacQueen G, Born L, Steiner M. The selective serotonin reuptake inhibitor sertraline: Its profile and use in psychiatric disorders. CNS Drug Reviews. 2001;**1**:1-24

[71] Taber GP, Pfisterer DM, Colbeg JC. A new and simplified process for preparing N-[4-(3,4-dichlorophenyl)-3,4-dihydro-1(2H)-naphthalenylidene]methanamine and a telescoped process for the synthesis of (1S-cis)-4-(3,4-dichlorophenol)-1,2,3,4-tetrahydro-N-methyl-1-naphthalenamine mandelate: Key intermediates in the synthesis of sertraline hydrochloride. Organic Process Research & Development. 2004;**8**:385-388

Section 4

Green Corrosion Inhibition

Chapter 6

Ionic Liquids as Green Corrosion Inhibitors for Industrial Metals and Alloys

Chandrabhan Verma, Eno E. Ebenso and
Mumtaz Ahmad Quraishi

Additional information is available at the end of the chapter

http://dx.doi.org/10.5772/intechopen.70421

Abstract

Present chapter describes recent advances in the field of development of ionic liquids as green and sustainable corrosion inhibitors for metals and alloys. The present chapter has been divided into several sections and subsections. Recently, development of the green and sustainable technologies for the corrosion prevention is highly desirable due to increasing ecological awareness and strict environmental regulations. In the last two decades, corrosion inhibition using ionic liquids has attracted considerable attention due to its interesting properties such as low volatility, non-inflammability, non-toxic nature, high thermal and chemical stability and high adorability. Several types of ionic liquids have been developed as "green corrosion inhibitors" for different metals and alloys such as mild steel, aluminum, copper, zinc, and magnesium in several electrolytic media. The ionic liquids are promising, noble, green and sustainable candidates to replace the traditional volatile corrosion inhibitors.

Keywords: ionic liquids, corrosion, adsorption, green corrosion inhibitors, designer solvents, ferrous and non-ferrous metals

1. Introduction

1.1. Corrosion and its economic impact

Corrosion is an irreversible and spontaneous deterioration of metal or alloy through chemical or electrochemical reaction with the environment [1, 2]. Corrosion causes enormous wastes of metallic materials which lead to enormous economic losses all over the world. Therefore, corrosion has drawn considerable academic and industrial attention [1–4]. According to highly cited study carried out by the National Association of Corrosion Engineers (NACE), in 1998, the total annual direct cost (estimated) of corrosion in U.S.A. was US $276 billion, equating approximately around 3.1% Gross Domestic Product (GDP; NACE 2002) [5]. In 2011, the total cost of corrosion in U.S.A. became more than US $2.2 trillion. As for as the corrosion cost in India is concern, it was around Rs. 2 lackscrores (US $45 billion) as proposed by 1st Global Corrosion Summit held in New Delhi, India in 2011 [6]. However, these estimated data are outdated and recently closer investigation of the NACE on the cost of corrosion is available according to which the annual global cost of corrosion is approximately US $2.5 trillion, equating 3.4% of the global GDP [7, 8]. In India, the annual corrosion cost is more than US $100 billion, while in South Africa, the direct corrosion cost is estimated to be around R130 billion (i.e. about US $ 9.6 billion) [7, 8]. There are several methods of corrosion protection have been developed such as coating, anodic and cathodic protections, alloying and de-alloying and use of synthetic corrosion inhibitors by suitably applying them we can reduce this cost of corrosion from 15% (US $375 billion) to 35% (US $ 875 billion).

1.2. Causes of corrosion

Pure metals are chemically unstable and undergo chemical and/or electrochemical reactions with their environments to form more stable oxides. The chemical reactivity of pure metals is related to their natural tendency of oxidation (except gold, silver and platinum), as they have tendency to return their natural state by chemical reactions with the constituents of environment [9–12]. Since corrosion is a spontaneous process, relative rate of corrosion among a given series of metals is related to the change in standard Gibb's free energy (ΔG^o). As more negative value of ΔG^o as high spontaneity of reaction and consequently higher corrosion rate [9–12]. When metals and alloys exposed to environment and particularly in acid solution during several industrial processes like acid pickling, acid descaling, etc., corrosion will undergo forming stable oxides [13–15]. Therefore, these processes required some additives known as corrosion inhibitors that form protective covering over the metallic surface and isolate metals from the environment and thereby inhibit the corrosive degradation [13–17]. The corrosion products such as rust and scale can also act as corrosion inhibitors by accumulation on the surface and act as physical protective barrier. The natural tendency of metallic corrosion can be affected by several factors, however, the relative rate of corrosion of any particular metal is depending upon the Pilling—Bedworth ratio which is defined as Md/nmD, where m and d are the atomic weight and density of the metal, respectively and M and D are the molecular weight and density of scale (corrosion product) accumulated on the metallic surface, and n denotes the number of metallic atoms in the molecular formula of corrosion product (rust or scale); for example for Fe_2O_3 and Al_2O_3, n = 2 [18, 19]. The magnitude of Pilling – Bedworth ratio can be used to explain where the surface film will be protective or not. The volume of corrosion product will be small than the volume of metal from which it was formed for $Md/nmD < 1$, in this situation it is expected that surface film of corrosion product contains pores and cracks that would be relatively non-protective. On the other hand, volume of corrosion product will be larger than the volume of metal for $Md/nmD > 1$, in that situation it is expected that surface film of corrosion product is relatively more compressed and compact and consequently the metal would be relatively more protected.

1.3. Corrosion prevention methods and corrosion inhibitors

There are several methods of corrosion protection have been developed among which, synthetic corrosion inhibitors are one of the best methods due to its advantages such as cost effectiveness and ease of application in industry [20–23]. The flow diagram of the available corrosion protection measures is shown in **Figure 1**. The passivating inhibitors are also known as anodic inhibitors because they general inhibit the metallic corrosion by forming the surface oxide (passive) film and causes the large anodic shift corrosion potential (E_{corr}) [24]. The passivating inhibitors can be further classified into oxidizing anions that passivate the metallic surface in the absence of oxygen such as chromate, nitrite and nitrate and non-oxidizing anions that can passivate the metallic surface only in the presence of oxygen such as phosphate, tungstate and molybdate. The cathodic inhibitors either decrease the rate of cathodic reactions or precipitate on the cathodic areas to increase the surface impedance that decrease the diffusion of reducible species to these areas [24]. The cathodic inhibitors act by three different mechanisms namely, cathodic poisons, cathodic precipitates and oxygen scavengers. Generally, arsenic and antimony make the association of hydrogen more difficult and act as cathodic poisons, calcium, zinc and magnesium precipitates in their oxide forms and act as cathodic precipitates and sodium sulfite and hydrazine react with surrounding oxygen and act as oxygen scavengers [25–28].

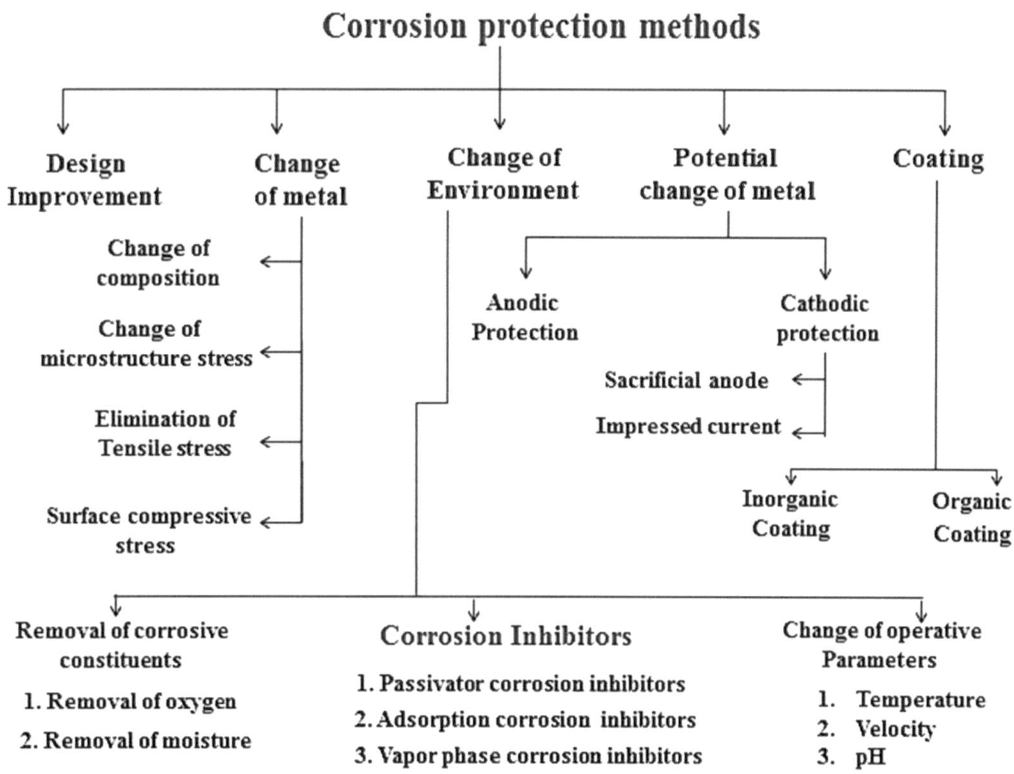

Figure 1. Available methods of metallic corrosion protection.

Organic compounds are also known as filming inhibitors; generally inhibit metallic corrosion by forming the protective surface film that isolates the metal form the surrounding (corrosive) environments. Most of the well know organic inhibitors are heterocyclic compounds containing polar functional groups such as $-NO_2$, -OH, $-OCH_3$, $-CH_3$, $-NH_2$, $-COOC_2H_5$, $-CONH_2$, -COOH, etc. [29–31]. These polar functional groups and conjugated π-electrons of multiple bonds (double and triple) act as adsorption centers during metal-inhibitor interactions. This type of adsorption results into blocking of anodic and cathodic reactions indirectly. The adsorption of these inhibitors is affected by several factors such as nature and magnitude of charge present on metal, nature of electrolyte, electronic structure of inhibitor molecules, nature of substituents, solution temperature, exposure time etc. [29–34].

1.4. Ionic liquids as green corrosion inhibitors

"Green chemistry" which is a relatively new and rapidly growing area of chemistry that involves designing of products and processes that reduce the use and production of toxic substances [35–38]. Recently, worldwide growing ecological awareness and strict environmental protocols do not permit the synthesis and utilization of hazardous traditional volatile corrosion inhibitors. Therefore, there is vital need for improvement in the synthetic and engineering chemistry either by environmental friendly starting materials or proper designing for synthesis using non-classical energy sources such as ultrasound and microwave heating. In this regard use of multi component reactions (MCRs) in combination with ultrasonic (sonochemical) and microwave irradiation is one of the best alternative synthetic strategies toward "green synthesis." Recently, scientists are trying to develop plant extracts and drugs as green corrosion inhibitors due to their natural and/or biological origins and non-toxic nature [39–41]. However, extraction and purification of plant extracts is very tedious, laborious, extremely expensive, time consuming and requires large amount of organic solvents [42, 43]. Therefore, there is need to develop "green inhibitors" by proper designing of the synthesis that can be achieved either by using cheap and environmental friendly starting materials or by synthesizing them from one step MCR reactions.

Toward, "green chemistry," utilization of ionic liquids has immersed as new strategy due to its several fascinating properties such as low melting point (lower that 100°C), high polarity, low toxicity, low vapor pressure, very high thermal and chemical stability, less hazardous influence on environment and living being [44–48]. By definition, ionic liquids are materials that mainly composed of ions with melting point below than 100°C. The properties of ionic liquids could be modified according to the need by proper selection of cations and anions, which is the greatest advantage for designing ionic liquids of specific properties [49–51]. Due to this reason ionic liquids are also known as "designer chemicals" that have potential to consume as solvent or catalysis for various chemical transformations [44–51]. The rapid utilization of ionic liquids in almost all fields of chemistry and chemical engineering is resulted to their above mentioned fascinating properties which enable them as "green and sustainable chemicals" having tendency to dissolve wide range of inorganic and organic compounds. The ionic liquids follow the principals of "green chemistry" proposed by Paul Anastas and John Warner [52–54].

http://dx.doi.org/10.5772/intechopen.70421

1.4.1. Properties and applications of ionic liquids

The ionic liquids have several fascinating properties such as low volatility (low vapour pressure), very high stability over wide range of pH and temperature, capability to dissolve a wide range of organic and inorganic compounds as they generally exist in their ionic forms through which they easily dissolve in polar solvents like H_2O, HCl, etc., moreover, their cationic counterparts generally contain large organic moieties through which they are capable to dissolve non-polar organic compounds, capability to solubilize gases like H_2, CO, CO_2 etc., dependency of solubility on the nature of cations and anions, acceleration of reaction rate for chemical transformation under microwave heating, long time stability without decomposition and their high selectivity [55–62]. These fascinating properties of ionic liquids make them good candidature to replace conventional organic volatile solvents with non-conventional ionic liquids that have been employed in variety of chemical transformations such as solvents for synthesis of nanomaterials and nanostructure, biochemical transformations, nucleophilic substitution reactions, electrodeposition of metals and semiconductors and solvent extraction, separation of petrochemical relevance mobile phase converter in HPLC, catalyst in various chemical and biochemical transformations, dye sensitizer for solar cells, oil shale processing, etc. (**Figure 2**) [55–62].

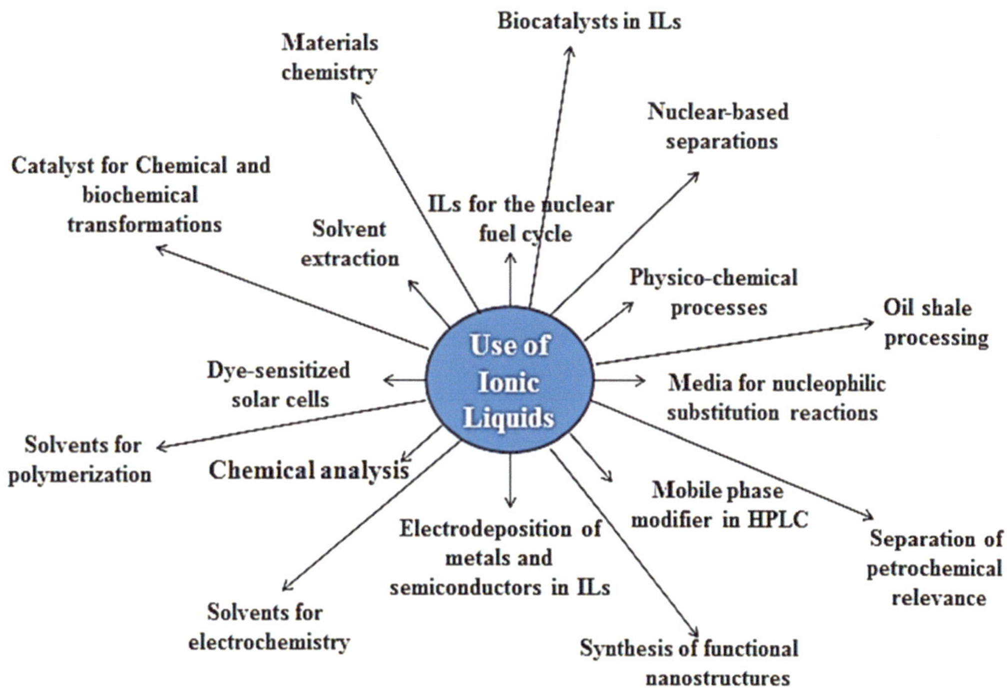

Figure 2. Applications of ionic liquids.

1.4.2. Classification of ionic liquids

The ionic liquids can be classified into several categories based on various bases. Hajipour and Refiee [63] have classified the ionic liquids into eleven classes namely, neutral ionic liquids, acid ionic liquids, basic ionic liquids, ionic liquids with amphoteric anions, functionalized ionic liquids, protic ionic liquids, chiral ionic liquids, supported ionic liquids, bio-ionic liquids, poly-ionic liquids, and energetic ionic liquids and also have described common features and properties of these ionic liquids. However, Suresh and Sandhu [62, 63] classified ionic liquids into only two classes namely, cationic and anionic ionic liquids. They were further subdivided anionic ionic liquids into several subclasses namely, borates, dicyanamide (DCN), Halide, Bis(trifluoromethylsulfonyl)imide (NTF), nonaflate (NON), phosphate, sulfate, sulfonate, thiocyanate (SCN), tricyanomethide (TCC) based anionic liquids. Some common classes of ionic liquids with examples and their salient features are described in **Table 1**.

SN	Types of ionic liquids (classes)	Some typical examples	Remark
1	Neutral ionic liquids	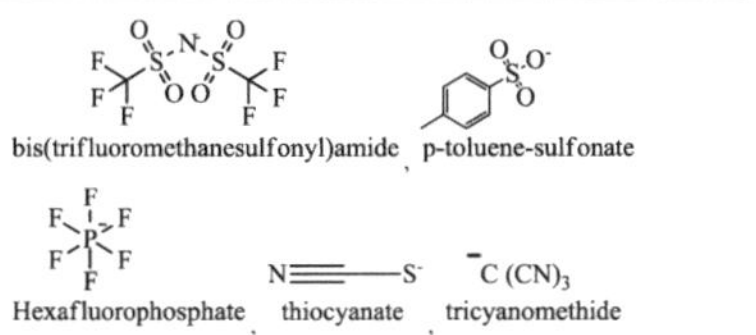	Anions are associated with cations with weal electrostatic interaction, low melting point, low viscosity, used as inert solvent, good thermal and electrochemical stability
2	Acidic ionic liquids	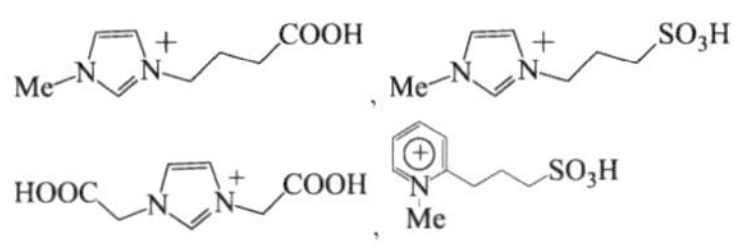 $Me_3\overset{+}{N}$-CH_2COOH Anions= (HSO_4^-, $H_2PO_4^-$, $M_3Cl_7^-$; Zn, Fe, Al)	Ionic liquids with acidic cations or acidic anion, enhanced solubility in water, possess good catalytic efficiency
3	Basic ionic liquids	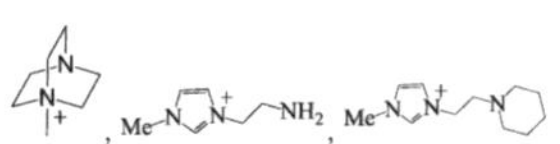 Common anions: Cl^-, CH_3OO^-, oxalate ion, etc.	These ionic liquids are basic in nature due to presence of one or more amine group (1°, 2° or 3° amine)
4	Functionalized ionic liquids	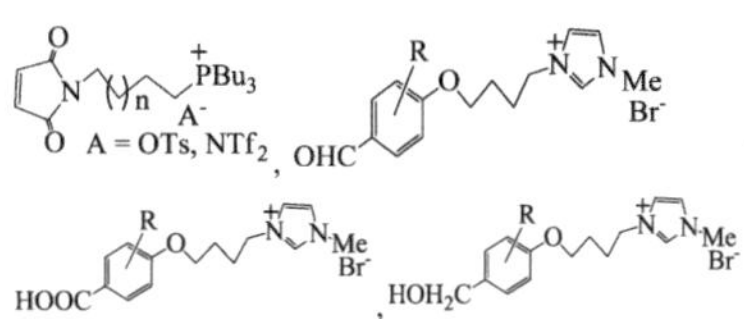	Ionic liquids that has a covalently bound functional group on the cation and/or anion
5	Supported ionic liquids	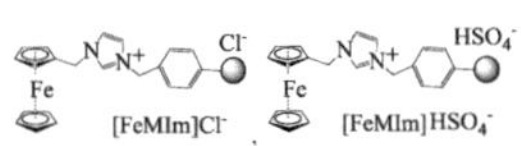	

Table 1. Classification of ionic liquids and their common features.

1.5. Comparison between organic inhibitors and ionic liquids

Over past two decades corrosion inhibition using ionic liquids (ILs) has experienced an outstanding growth and abundant examples on corrosion inhibitions are available that have been effectively carried out in different corrosive media. Although, traditional volatile compounds have been most extensively used as corrosion inhibitors in several industries. However, most of them are toxic for living being and environment [64–66]. In view of this, ionic liquids (ILs) have been used extensively in recent years. Ionic liquids have several advantageous physiochemical properties including non-toxic, high conductivity, non-flammability, as well as high thermal and chemical stability [35–63]. One of the most significant characteristics of ionic liquids is their environmental friendly and non-hazardous nature due to their non-negligible vapour pressure. Unlike to traditional volatile corrosion inhibitors, due to their extremely low vapour pressure these compounds will not evaporate and will not contaminate the surrounding environment [67, 68]. Additionally, sometimes the use of organic inhibitors particularly polymeric and high molecular weighted organic compounds is limited due to their extremely low solubility in the polar corrosive media [69–72]. However, ionic liquids are highly soluble in the polar corrosive environments due to their ionic nature [73]. Furthermore, there is limit less prospect of suitably modifying the structure of the anion and cation of any given ionic liquids delivers an unlimited amount of potential derivatives having numerous physiochemical properties, while this type of modification is not possible with volatile corrosion inhibitors. In summary, the use of ionic liquids as corrosion inhibitors is preferred as compared to traditional volatile (toxic) corrosion inhibitors due to their several advantageous physiochemical properties including their high solubility, non-toxic, high conductivity, non-flammability, less volatility as well as high chemical stability and more importantly due to their "green and sustainable" nature.

2. Applications of ionic liquids as corrosion inhibitors

Several fascinating properties of the ionic liquids make them ideal candidates to replace the traditional corrosion inhibitors that have several adverse effects on environment and living beings. Recently, a large number of works have been reported describing the use of ionic liquids as corrosion inhibitors.

2.1. Ionic liquids as corrosion inhibitors for mild steel

Mild steel is most frequently used as constructional material for several industries due to its high mechanical strength and low cost [74, 75]. However, these materials are highly reactive and undergo corrosive degradation during various industrial processes like acid cleaning, acid descaling, acid etching, and acid pickling processes that require use of additives in order to increase the lifespan of metal/alloy has used [76]. The use of organic compounds containing heterocyclic rings and polar fictional groups such as amino, hydroxyl, methyl, methoxy, nitro, nitrile, etc., as additive is one of the most important alterative to protect metals and alloys from these unsolicited reactions [74, 75]. These compounds inhibit corrosion by adsorbing over the metallic surface [74–77]. However, the use of these highly volatile traditional toxic corrosion inhibitors is limited due to increasing ecological awareness and strict environmental regulations. In this regards consumption of "ionic liquids" as corrosion inhibitors has become an important green alternative methods of corrosion protection. Literature survey reveals that several synthetic ionic liquids have been used as effective corrosion inhibitors for mild steel (or carbon steel) in various electrolytic media. Likhanova et al. [78] synthesized two ionic liquids namely, 1,3-dioctadecylimidazolium bromide ($ImDC_{18}Br$) and N-Octadecylpyridiniumbromide ($PyC_{18}Br$) using conventional and microwave heating methods, respectively and investigated their inhibition performance on mild steel corrosion in 1M H_2SO_4 using several experimental techniques. They were observed that studied ionic liquids acted as good corrosion inhibitors for mild steel in aqueous acid solution. The adsorption on metallic surface takes place via chemisorption mechanism which obeyed the Langmuir adsorption isotherm. Potentiodynamic polarization results revealed that applied ionic liquids behaved as mixed type inhibitors. These authors were proposed a mechanism of corrosion inhibition on the basis of results obtained from SEM-EDX, XRD and Mossbauer analyses. The inhibition performance of the 1-ethyl-3-methylimidazolium dicyanamide (EMID) on mild steel corrosion in 0.1M H_2SO_4 using several experimental techniques [79] has been tested. Results showed that EMID inhabits metallic corrosion by adsorption on the metallic surface which was confirmed by decreased values of C_{dl} and increased surface coverage in presence of the inhibitor. The adsorption of the EMID over metallic surface obeyed the Langmuir adsorption isotherm. The inhibition performance of two ionic liquids namely 1-butyl-3-methylimidazolium chlorides (BMIC) and 1-butyl-3-methylimidazolium hydrogen sulfate ([BMIM]HSO_4) on mild steel corrosion in 1M HCl have been studied by Zhang and Hua [80] using electrochemical and weight loss experiments. Results showed that the inhibition efficiency of both ionic liquids obeyed the order: ([BMIM]HSO_4) > (BMIC). They were found that adsorption of these compounds on mild steel surface obeyed the Langmuir adsorption isotherm. Potentiodynamic study suggested that both ionic liquids acted as mixed type inhibitors. The effect of temperature (303–333 K) was also investigated on both the ionic liquids. Finally, several activation and thermodynamic parameters such as energy of activation (E_a), enthalpy of activation (ΔH), entropy of activation (ΔS), adsorption constant (K_{ads}) and Gibb's standard free energy (ΔG^{o}) were calculated in order to explain the mechanism of adsorption and corrosion inhibition of both the ionic liquids.

The inhibition performance of 1-octyl-3-methylimidazolium bromide ([OMIM]Br) and 1-allyl-3-octylimidazolium bromide ([AOIM]Br) on mild steel corrosion in 0.5 M H_2SO_4 using weight loss, electrochemical, scanning electron microscopy (SEM) and Quantum chemical calculations techniques showed that both the ionic liquids acted as good corrosion inhibitors and their adsorption on the metallic surface obeyed the El-Awady thermodynamic–kinetic model and acted as slightly cathodic type inhibitors [81].

Table 2 represents the corrosion inhibition properties of several other ionic liquids that have been employed as inhibitors for mild steel corrosion in electrolytic media [82–116]. The chitosan-based ionic liquid was synthesized using oleic acid and p-toluene sulfonic acid and its corrosion inhibition efficiency was determined using several electrochemical measurements [117]. Results of the investigated study revealed that presence of the ionic liquid in the chloride containing corrosive medium decreased the rate of metallic dissolution as well as hydrogen evolution. Adsorption of the ionic liquid followed the Langmuir adsorption

Synthetic scheme and/or chemical structure of ionic liquids	Techniques	Nature of adsorption	Electrolytic media	Ref.
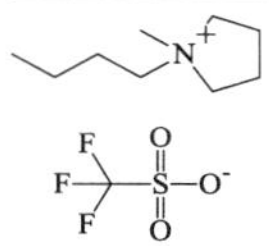	Electrochemical and scanning electron microscopy	Langmuir adsorption isotherm, mixed type	3.5% NaCl	[82]
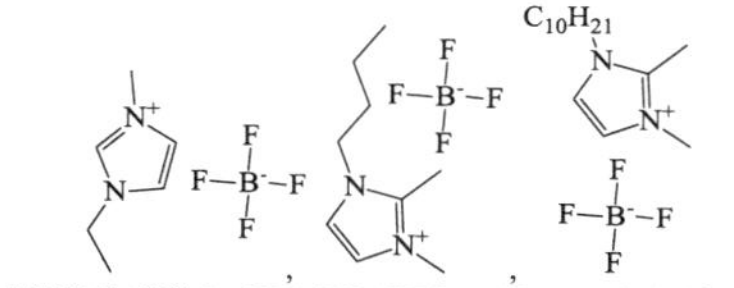 [EMIM]+[BF₄]−, [BDMIM]+[BF₄] −, [C10MIM]+[BF₄]−	Experimental, quantum chemical, Monte Carlo simulation	Langmuir adsorption isotherm, mixed type	1M HCl	[83, 84]
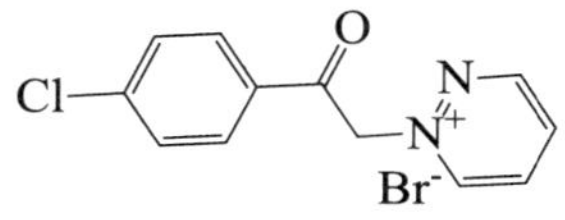 (CPEPB)	Gravimetric, electrochemical, quantum chemical calculations	Langmuir adsorption isotherm, mixed type	1M HCl	[85]
Scheme 1 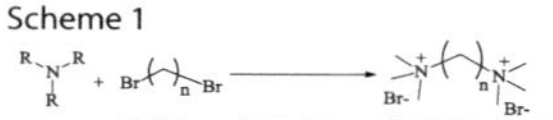 (G_2IL): n = 2; (G_3IL): n = 3; (G_6IL): n = 6	Gravimetric, electrochemical	Langmuir adsorption isotherm, mixed type	1M HCl	[86]
Scheme 2 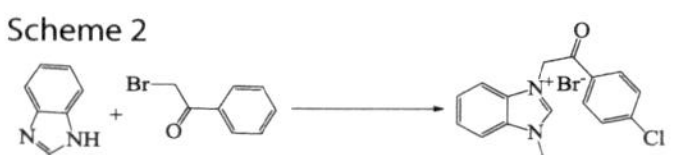 ([BMMB]+Br-)	Electrochemical, Quantum chemical calculations (DFT)	Langmuir adsorption isotherm, mixed type	1M HCl	[87]
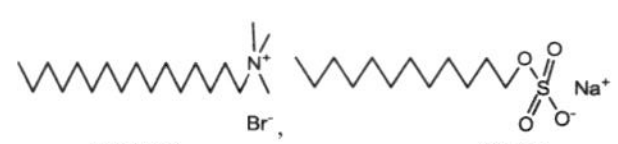 (CTAB) (SDS)	Electrochemical, Scanning electron microscopy	Flory–Huggins adsorption isotherm, mixed type	3.5% NaCl	[88]

Synthetic scheme and/or chemical structure of ionic liquids	Techniques	Nature of adsorption	Electrolytic media	Ref.
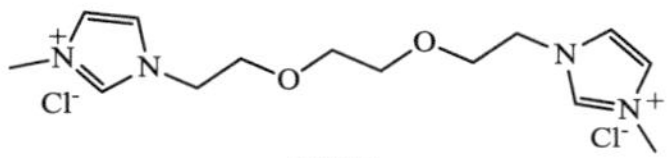 (TSIL)	Weight loss, electrochemical, SEM, AFM, contact angle method	Langmuir adsorption isotherm, mixed type	1M HCl	[89]
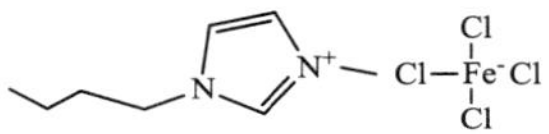 ($[C_4C_1im][FeCl_4]$)	Weight loss, electrochemical, SEM, DFT methods	–	Open and controlled environments	[90]
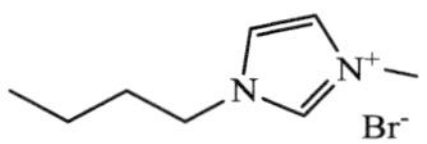 [BMIM]Br	Weight loss, electrochemical	Langmuir adsorption isotherm, mixed type	1M HCl	[91, 92]
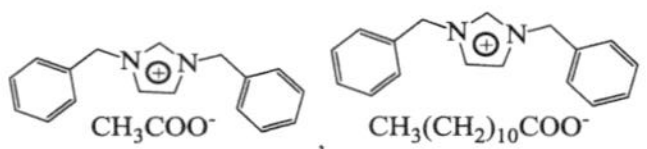 (DBImL), (DBImA)	Electrochemical, SEM	Langmuir adsorption isotherm, mixed type	1M HCl, 1M H_2SO_4	[93]
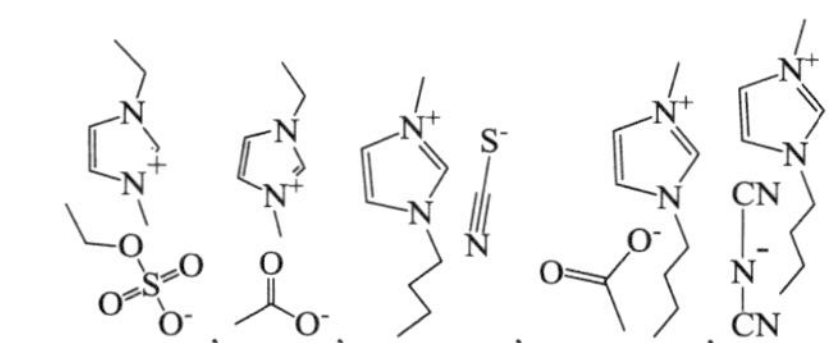 [EMIM] + $[EtSO_4]$ - , [EMIM] + [Ac] –, [BMIM] + [SCN] –, [BMIM] + [Ac] –, [BMIM] + [DCA] –	Electrochemical, spectroscopic, SEM, DFT, QSAR and Monte Carlo simulation	Langmuir adsorption isotherm, mixed type	1M HCl,	[94]
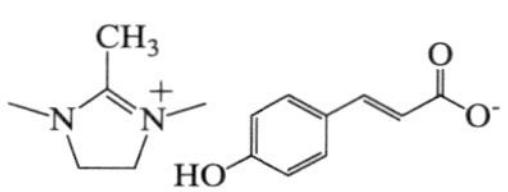	Electrochemical, Immersion, SEM	–	0.01M NaCl	[95]

Synthetic scheme and/or chemical structure of ionic liquids	Techniques	Nature of adsorption	Electrolytic media	Ref.
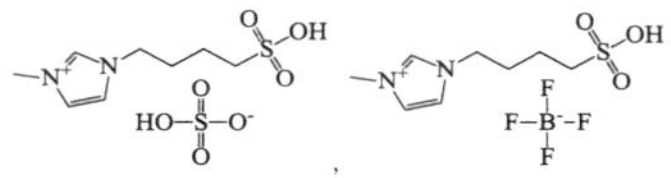 ([BsMIM]-[HSO$_4$]), ([BsMIM][BF$_4$]),	Gravimetric, electrochemical, SEM	Langmuir adsorption isotherm	1M H_2SO_4	[96]
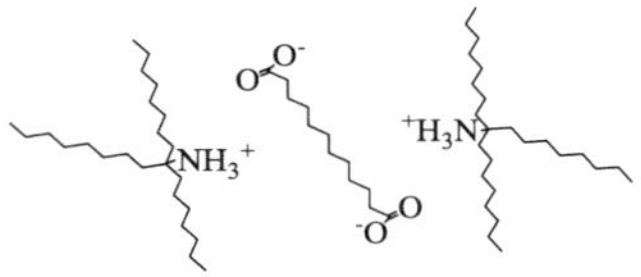	Weight loss and polarization techniques	Langmuir adsorption isotherm, mixed	Production water	[97]
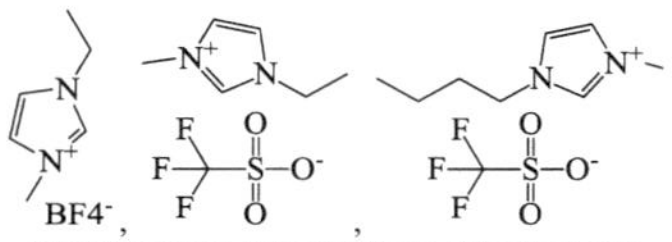 [EMIM][BF$_4$], [BMIM][Otf], [EMIM][Otf]	Electrochemical techniques	–	CO_2	[98]
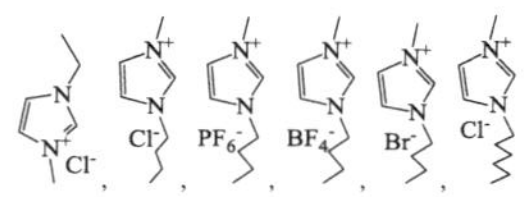 (EMIm Cl), (BMIm Cl), (BMIm PF$_6$), (BMIm BF$_4$), (BMIm Br), (HMIm Cl)	Electrochemical), AFM, dynamic light scattering (DLS), FT-IR and DFT	Langmuir adsorption isotherm, mixed type	2 M HCl	[99]
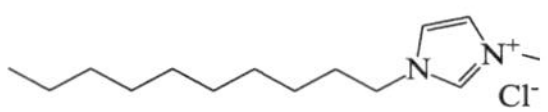 (DMICL)	Electrochemical, surface analysis techniques	–	NaCl (pH 3.8 and pH 6.8)	[100]
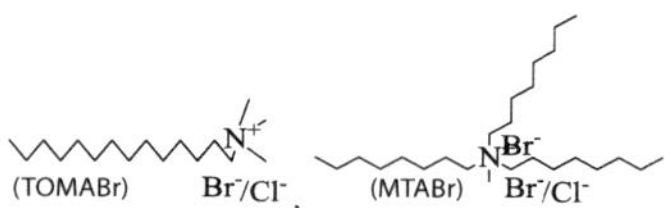	Electrochemical measurements	Flory-Huggins adsorption isotherm, mixed type	2M H_2SO_4	[101]

Synthetic scheme and/or chemical structure of ionic liquids	Techniques	Nature of adsorption	Electrolytic media	Ref.
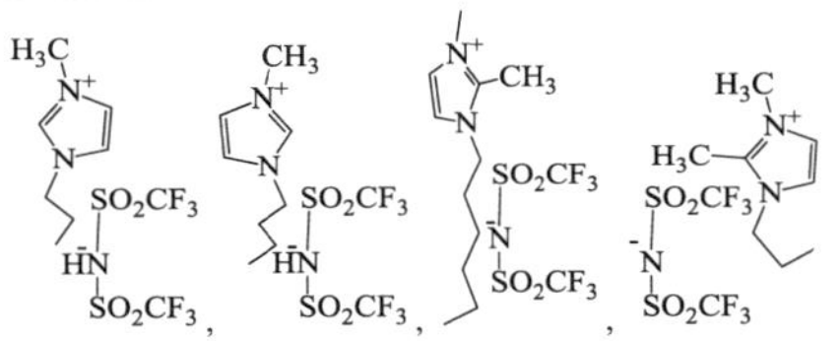	Weight loss, electrochemical measurements, QSAR, quantum chemical calculations	Langmuir adsorption isotherm, mixed type	1M HCl	[102]
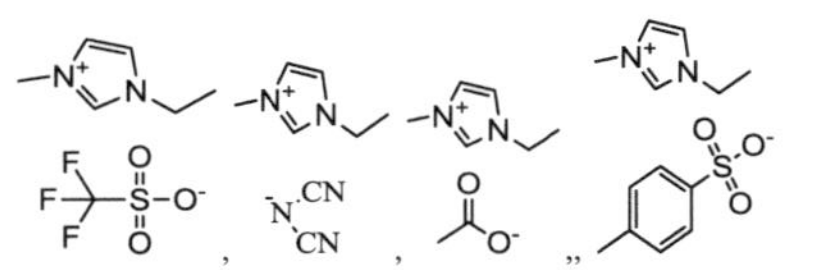 [emim]-[Otf], [emim]-[DCA], [emim][acetate], [emim][tosylate])	Electrochemical	–	CO_2 capture system	[103]
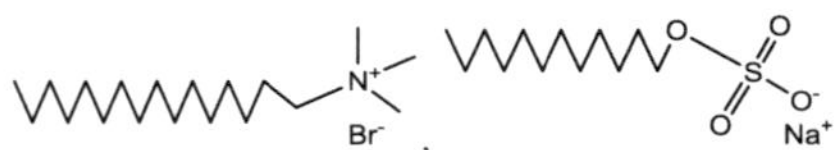 (CTAB), (SDS)	Electrochemical and surface analysis	Flory – Hugginsadsortion isotherm, mixed type	2 M HCl	[104]
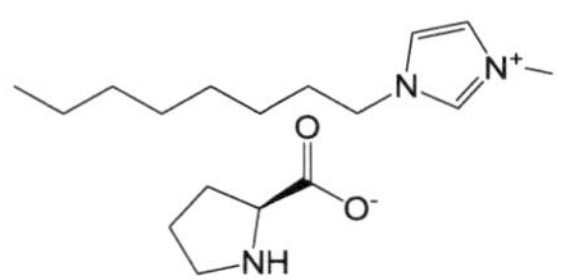	Weight loss, electrochemical, SEM, and quantum chemical calculation	Flory – Huggins isotherm, mixed type	0.5 M H_2SO_4	[105]
Scheme 5 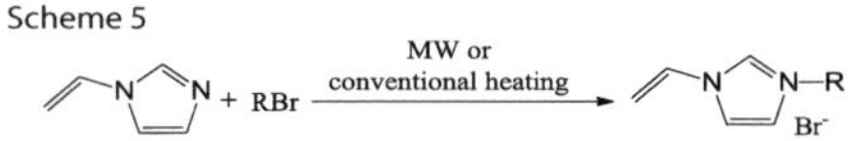 R= IL1:C_4H_9; IL2:C_8H_{17}; IL3:$C_{12}H_{25}$; IL4:$C_{18}H_{37}$; IL5: $C_{22}H_{45}$	Weight loss, electrochemical, SEM, AFm	Langmuir isotherm, mixed type but IL3 behave as cathodic type	1M H_2SO_4	[106]

Synthetic scheme and/or chemical structure of ionic liquids	Techniques	Nature of adsorption	Electrolytic media	Ref.
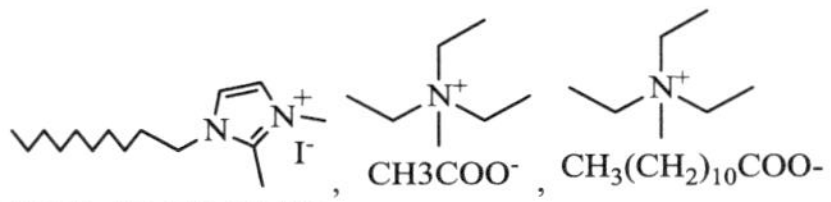 (DDI), (TMA) (TML)	Weight loss, electrochemical, SEM	Langmuir isotherm, mixed type	1M H_2SO_4	[107]
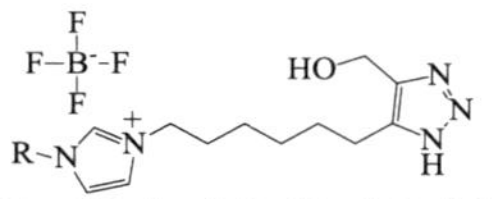 R= $-CH_3$ (I); $-C_4H_8$ (II); $-C_8H_9$ (III)	Electrochemical polarization test, SEM	–	Ethanol solution	[108]
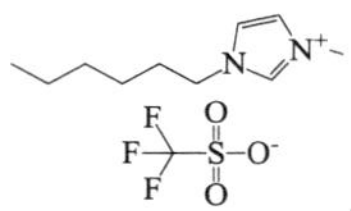 [HMIM][TfO], 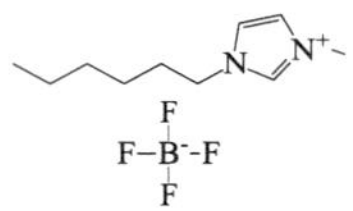 [HMIM][BF_4], 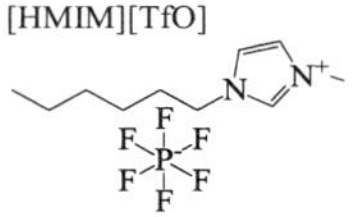 [HMIM][PF_6], 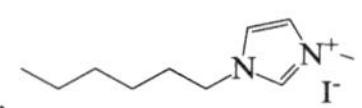 [HMIM][I]	Electrochemical, spectroscopic analyses, quantum chemical calculations	Langmuir isotherm, mixed type	1M HCl	[109]
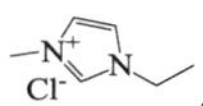 ([EMIm]Cl),  ([Py1,4]Cl)	Weight loss, electrochemical	–	Arabian Gulf Sea-water	[110]
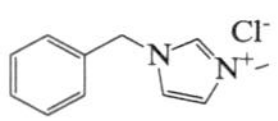 (BMIC)	Weight loss, electrochemical	Langmuir adsorption isotherm, mixed type	2M H_2SO_4 and 3.5% NaCl	[111]

Synthetic scheme and/or chemical structure of ionic liquids	Techniques	Nature of adsorption	Electrolytic media	Ref.
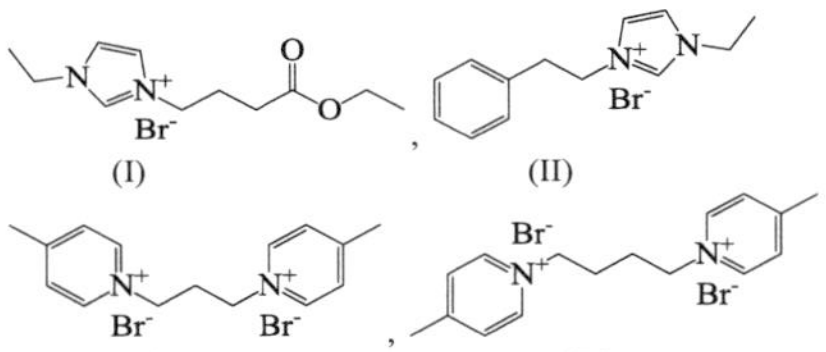 (I), (II), (III), (IV)	Potentiodynamic polarization, linear polarization and weight loss	Langmuir adsorption isotherm, mixed type	1M HCl	[112]
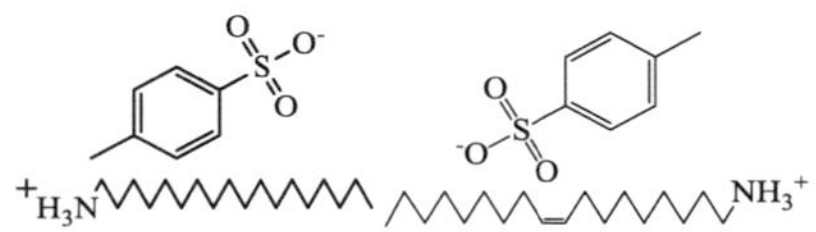 (ODA-TS) (OA-TS)	Electrochemical, SEM, EDX, contact angle measurement	Langmuir adsorption isotherm, mixed type	1M HCl	[113]
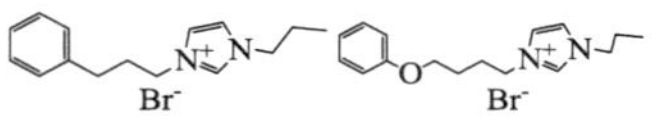 (PPIB1) (PPIB4)	Weight loss and electrochemical methods	Langmuir adsorption isotherm, mixed type	1M HCl	[114]
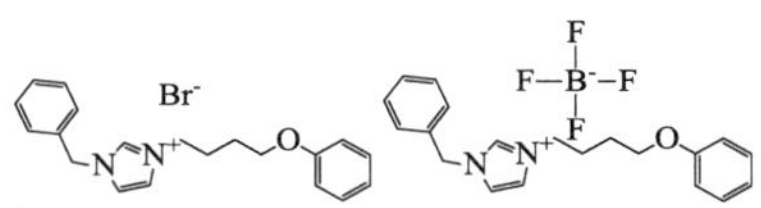 (MA1) (MA2)	Weight loss and DFT studies	Langmuir adsorption isotherm	1M HCl	[115]
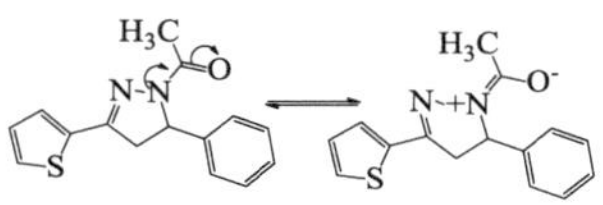	Weight loss and electrochemical methods	Langmuir adsorption isotherm, mixed type	1M HCl/1M H_2SO_4	[116]

Synthetic scheme and/or chemical structure of ionic liquids	Techniques	Nature of adsorption	Electrolytic media	Ref.
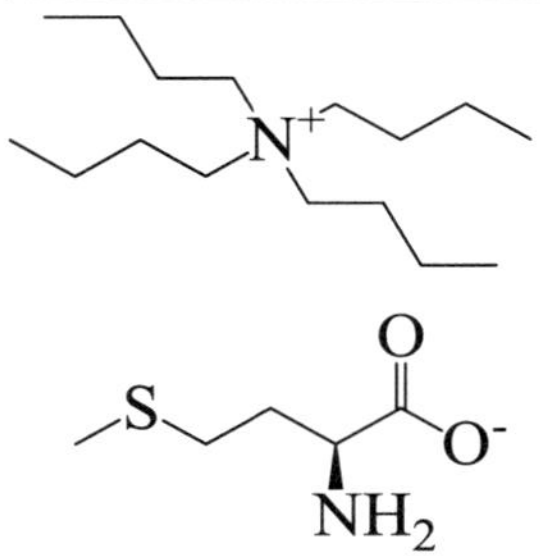	Electrochemical, Quantum, surface analysis methods	Mixed type	1M HCl	[121]
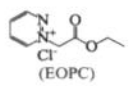	Weight loss, electrochemical methods	Langmuir adsorption isotherm, mixed type	0.5M H_2SO_4	[122]
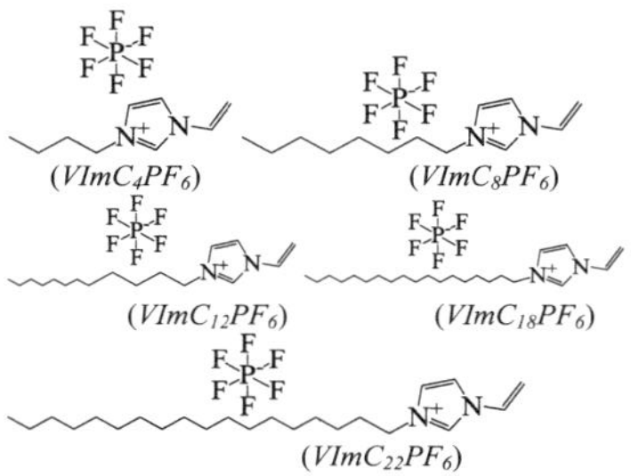	Weight loss and electrochemical polarization methods	Langmuir adsorption isotherm, mixed type	1M H_2SO_4	[123]

Table 2. Ionic liquids as corrosion inhibitors for mild steel in different electrolytic media, their mode of adsorption and techniques used for evaluation of the inhibition performance.

isotherm. Polarization study suggested that investigated ionic liquid acted as mixed type inhibitor. Tseng and coworkers [118] investigated the corrosion characteristics of carbon steel, 304 stainless steel (304 SS) and pure titanium (Ti) in aluminum chloride–1-ethyl-3-methylimidazolium chloride ionic liquid for the first time. These authors reported the active-to-passive transition behavior for CS sample. Among the tested materials 304 SS exhibited the maximum stability in the high chloride environment. The most peculiar finding was that Ti was severally corroded in the ionic liquid because it does not undergo passivation. The ionic liquid in non-aqueous, low-oxygen and high halogen containing showed different corrosion behavior and mechanism. Similar observation has been reported by other authors for different metals including copper, nickel and stainless steel [119]. Recently, the inhibition behavior of 1,4-di [1-methylene-3-methyl imidazolium bromide]- benzene on mild steel corrosion in 1M H_2SO_4 have been studied using electrochemical and surface analysis methods [120]. The ionic liquid under taken in the study inhibits metallic corrosion by adsorbing on the surface which mechanism obeyed the Langmuir adsorption isotherm. The adsorption mechanism was supported by SEM, EDX and AFM analyses. Polarization study reveals that studied ionic liquid acted as mixed type inhibitor. The ongoing discussion reveals that although, several classes of ionic liquids have been used as effective inhibitors for mild steel corrosion in various aggressive media, however, imidazole based ionic liquids have been used most extensively [78–84, 89–96, 98–112, 114–116, 123].

2.2. Ionic liquids as corrosion inhibitors for aluminum

Aluminum is the second most commonly used metal due to its several fascinating properties like its low atomic mass and negligible standard electrode potential. Several traditional organic and inorganic compounds have been used previously in order to protect dissolution of protective surface oxide film and ultimately decrease the corrosion rate. However, employment of the ionic liquids as corrosion inhibitors is limited as literature survey reveals that only few works are available describing the corrosion inhibition performance of ionic liquids. The inhibition performance of 1-butyl-3-methylimidazoliumchlorides (BMIC), 1-hexyl-3-methylimidazolium chlorides (HMIC) and 1-octyl-3-methylimidazoliumchlorides (OMIC) on aluminum corrosion in 1M HCl using electrochemical and weight loss methods showed that inhibition efficiencies of these ionic liquids increase with increasing their concentration and obeyed the order: OMIC > HMIC > BMIC [124]. Potentiodynamic study revealed that all ionic liquids acted as mixed type inhibitors and their adsorption on aluminum surface followed the Langmuir adsorption isotherm. The inhibition efficiency of an ecofriendly ionic liquid, 1,3-bis (2-oxo-2-phenylethyl)-1H-imidazol-3-ium bromide (OPEIB) on 6061 Al-15 alloy in 0.1 M H_2SO_4 solution using electrochemical impedance spectroscopy and potentiodynamic polarization, scanning electron microscopy (SEM) and energy dispersive X-ray (EDX) spectroscopic methods revealed that it is a good corrosion inhibitor and its adsorption on aluminum surface obeyed the Temkin adsorption isotherm [125]. The three synthesized ILs, namely poly(ionic liquid)s (PILs), namely (poly(1-vinyl-3-dodecyl-imidazolium) (PImC12), poly(1-vinyl-3-octylimidazolium) (PImC8) and poly(1-vinyl-3-butylimidazolium) (PImC4) hexafluorophosphate) tested as inhibitor for aluminum alloy AA6061 in 0.1-1.0 M H_2SO_4 solution [126]. Results showed that they act as mixed type inhibitor and their inhibition efficiencies obeyed the order: (PImC12 > PImC8 > PImC4). Adsorption of these ionic liquids followed the Langmuir adsorption isotherm. Four newly synthesized quaternary ammonium based surfactants in the series of hexanediyl-1,6-bis-(diethyl alkyl ammonium bromide), designated as $C_mC_6C_m(Et)\cdot 2Br$ (m = 10, 12, 14, 16), were synthesized and evaluated as inhibitors for aluminum corrosion in 1M HCl solution [127]. Results showed that all investigated surfactants act as good inhibitors and inhibit corrosion by becoming adsorbate at metal/electrolyte interfaces and their adsorption on metallic surface obeyed the Langmuir adsorption isotherm. Trombetta et al. [128] studied the stability of the aluminum in 1-butyl-3methylimidazolium tetrafluroborate ionic liquid and ethylene glycol mixtures using electrochemical impedance spectroscope (EIS). These authors observed decrease in polarization resistance and increase in the capacitance related with the passive oxide dielectric properties on increasing the ethylene glycol and/or water content in the mixtures. Presence of salts namely $Na_2B_4O_7.7H_2O$ and NaH_2PO_4 in the mixtures, stabilize the oxide payer form over the metallic surface and thereby reduce the changes of metallic corrosion. The inhibition behavior of 1,3-bis(2-oxo-2-phenylethyl)-1H-imidazol-3-ium bromide (OPEIB) on 6061 Al-15 vol. pct. SiC(p) composite in 0.1M H_2SO_4 solution was studied by Shetty and Shetty [125] using electrochemical (EIS and PDP), SEM and EDX methods. The investigated ionic liquid exhibits the maximum efficiencies of 96.7 and 94% using PDP and EIS methods, respectively. Potentiodynamic polarization study further reveals that studied ionic liquid behaves as cathodic type inhibitor and its adsorption on the composite surface followed the Temkin adsorption. Li et al. [129] study the inhibition behavior of tetradecylpyridinium bromide (TDPB) on aluminum corrosion in 1M HCl solution using weight loss and electrochemical methods. Results of the investigation showed that TDPB inhibits the aluminum corrosion by adsorbing on the metallic surface. The adsorption of the TDPB followed the Langmuir adsorption isotherm. Polarization study suggested that TDPB acts as cathodic type inhibitor for acidic aluminum corrosion. Bermudez and coworkers [130] investigated the surface interactions of seven alkylimidazolium ionic liquids with aluminum alloy Al 2011 using immersion test. The immersion experiments for aluminum corrosion was carried out in 1 and 5 wt.% of 1-ethyl,3-methylimidazolium tetrafluoroborate (IL1) in water. Results showed that neat solution of ionic liquids did not cause any corrosion. The inhibition behavior was discussed on the basis of SEM, EDX, XPS and XRD techniques.

2.3. Ionic liquids as corrosion inhibitors for copper and zinc

Copper and its alloys have been extensively employed in industries for various applications such as building construction, electricity, electronics, coinages, ornamental and formation of industrial equipment due to their relatively good thermal, electrical, mechanical and corrosion resistance properties [131]. However, in presence of aggressive anions like chloride, sulphate and nitrate these materials undergo sever attack resulting into loss of these materials due to corrosion occurs [132, 133]. Similar to the aluminum the use of ionic liquids as corrosion inhibitors for copper and zinc is also limited as literature survey revealed that only few ionic liquids have been used as corrosion inhibitors for these materials. Qi-Bo and Yi-Xin [134] newly synthesized three ionic liquids namely 1-butyl-3-methylimidazolium hydrogen sulfate ([BMIM]HSO_4), 1-hexyl-3-methylimidazolium hydrogen sulfate ([HMIM]HSO_4), and 1-octyl-3-methylimidazolium hydrogen sulfate ([OMIM]HSO_4) and studied their inhibition efficiency on copper corrosion in 0.5 M H_2SO_4 using electrochemical impedance spectroscopy and potentiodynamic polarization techniques. The inhibition efficiency of the ionic liquids follows the order: [OMIM]HSO_4 > [HMIM]HSO_4 > [BMIM]HSO_4. Results obtained by these authors showed that adsorption of the studied ionic liquids followed the Langmuir adsorption isotherm. Polarization study revealed that these ionic liquids behaved as mixed type inhibitors. Gabler et al. [135] studied the inhibition performance of two ionic liquids namely (2-hydroxyethyl)-trimethyl-ammonium (IL1) and Butyl-trimethyl-ammonium (IL2) with identical anions; bis(trifluoromethyl-sulfonyl)imide on $CuSn_8P$ and steel 100Cr_6, purchased from Metal Supermarkets (Brunn am Gebirge, Austria) using inductively coupled plasma optical emission spectrometry (ICP-OES), scanning electron microscopy (SEM) with energy dispersive X-ray spectrometry (EDX) and X-ray photoelectron spectroscopy (XPS) in water in the absence and presence of 1.5% of the ionic liquids. Manamela et al. [136] studied the inhibition performance of two ionic liquids; 1-butyl-3-methylimidazolium tetrafluoroborate [BMIM][BF_4^-] and 1-decyl-3-methylimidazolium tetrafluoroborate [DMIM][BF_4^-] on corrosion of zinc in 1M HCl using gravimetric analysis and theoretical Density Functional Theory (DFT) approach, using the B3LYP functional. Results showed that both the ionic liquids acted as good corrosion inhibitors and their inhibition efficiencies increase with increasing their concentrations. The inhibition efficiencies of the ionic liquids obeyed the order: [DMIM][BF_4^-] > [BMIM][BF_4^-]. Values of activation energy (E_a) and enthalpy of activation (ΔH) suggested that both the ionic liquids adsorbed over the surface through physisorption mechanism. Adsorption of these ionic liquids on metallic surface followed the Langmuir adsorption isotherm.

2.4. Ionic liquids as corrosion inhibitors for magnesium

Unlike active light metals such as aluminum and titanium, magnesium based alloys do not form protective passivating film. Moreover, these alloys easily react with the components of environment to from hydroxides, oxides, carbonates films that are highly porous, inhomogeneous and poorly bonded that cannot provide satisfactory protection to the metals against corrosion. Among the available methods of corrosion protection, organic coating is one of the best methods. Huanga et al. [137] has presented an early review on the corrosion protection of magnesium by some ionic liquids. However, present chapter is describing the few recent advances in the utilization of ionic liquids as corrosion inhibitors. Suna et al. [138] have investigated the inhibition effect of six phosphonium cation based ionic liquids (ILs) namely, tetradecyltrihexylphosphonium diphenylphosphate (1), tetradecyltrihexylphosphoniumdibutylphosphate (2), tetradecyltrihexylphosphonium bis(2-ethylhexyl) phosphate (3), tetradecyltrihexyl phosphonium diisobutyldithiophosphinate (4), tetradecyltrihexylphosphoniumbis(2,4,4-trimethyl pentyl) phosphonate (5), and tetradecyltrihexyl phosphonium O,O-diethyl dithiophosphate on magnesium alloys using electrochemical and surface investigation methods.

3. Ionic liquids as corrosion inhibitors: DFT study

Nowadays, several computational methods particularly, DFT (Density Functional Theory) based quantum chemical calculations have been emerged as potential tools for studying the interactions between inhibitors and metallic surface. The DFT calculations provide several important parameters such as energies of highest occupied molecular orbital (E_{HOMO}), lowest unoccupied molecular orbital (E_{LUMO}), energy band gap ($E_{LUMO} - E_{HOMO} = \Delta E$), global electronegativity (χ), global hardness (η) and softness (σ), fraction of electron transfer (ΔN) and dipole moment (μ). In general, value of E_{HOMO} is related with electron donating ability, while the value of E_{LUMO} related with the electron accepting ability of the inhibitor molecules [74–77]. A higher value of E_{HOMO} and lower value of E_{LUMO} associated with high inhibition performance. The inhibition efficiency of inhibitor increases with decreasing the energy band gap (ΔE). A high value of global electronegativity (χ) is related with lower electron donating ability and therefore, the value of electronegativity (χ) inversely related with the inhibition efficiency order [74–77]. Inhibition efficiency of the inhibitor molecules decreases with increasing the hardness (η) and decreasing the softness (σ). Generally, inhibition performance of the inhibitor molecules increases with increasing their dipole moment (μ), however, negative trends of the inhibition efficiency is also reported by several authors [74–77]. Lastly, the value of electron transfer gives direct information about the relative extent of metal-inhibitor interactions. A high value of ΔN is associated with high charge transfer and therefore high inhibition efficiency [74–77, 102].

The DFT based quantum chemical calculations have also been employed to describe the adsorption behavior of some ionic liquids on the metallic surface. Our research group [102] studied the adsorption behavior of four imidazolium-based ionic liquids, namely 1-propyl-3-methylimidazolium bis(trifluoromethyl-sulfonyl) imide ([PMIM][NTf2), 1-butyl-3-methylimidazoliumbis(trifluoromethyl-sulfonyl) imide ([BMIM][NTf2), 1-hexyl-3-methylimidazolium bis (trifluoromethyl-sulfonyl) imide([HMIM][NTf2]), and 1-propyl-2,3-methylimidazolium bis (trifluoromethyl-sulfonyl) imide ([PDMIM][NTf2]) on mild steel corrosion in 1M HCl using experimental and quantum chemical calculations. The inhibition efficiencies of these ionic liquids follow the experimental trend: [PDMIM][NTf2] > [HMIM][NTf2] > [BMIM][NTf2] > [PMIM][NTf2]. The values of E_{HOMO} and E_{LUMO} are well satisfied the experimental order of inhibition efficiency. Results showed that [PDMIM][NTf2] exhibited the lowest value of ΔE and therefore related with the highest chemical reactivity and inhibition efficiency. The values of dipole moment (μ) and the molecular volume (MV) did not show any regular trends. However, the values of global softness (σ) again show that the [PDMIM][NTf2] is most soft molecule among the tested compounds thereby associated with highest chemical reactivity and inhibition efficiency. The quantum chemical calculations provide good insight about the inhibition mechanism and well supported the experimental order of inhibition efficiency. Similar observations were reported for few other metals and alloys in several corrosive media [82, 139–143].

4. Mechanism of corrosion inhibition

Similar to most of the organic corrosion inhibitors, ionic liquids (ILs) inhibit metallic corrosion by blocking the anodic and cathodic sites present over the metallic surface [78, 144, 145]. Therefore, inhibition of metallic corrosion in presence of ionic liquids involves blocking of anodic oxidative metallic dissolution as well as cathodic hydrogen evolution reactions [78, 144]. The mechanism of metallic (M) corrosion inhibition by ionic liquids in sulphuric acid has been described below. The inhibition mechanism of metallic corrosion by ionic liquids in other protic acidic solutions such as in HCl and HNO_3 will be similar because of their similar nature. The only difference in their nature is that they possess different counter ions (Cl^-, NO_3^-) rather than sulphate ion of sulphuric acid. According to Likhanova et al. [78], anodic dissolution of metals (M) in aqueous acidic solution (e.g. H_2SO_4) can be represented as follows [78]:

$$M + nH_2O \longleftrightarrow M(H_2O)n_{ads} \tag{1}$$

$$M(H_2O)n_{ads} + SO_4{}^{2-} \longleftrightarrow M\left[(H_2O)_n SO_4^{2-}\right]_{ads} \tag{2}$$

$$M\left[(H_2O)_n SO_4^{2-}\right]_{ads} \longrightarrow M\left[(H_2O)_n SO_4\right]_{ads} + 2e^- \tag{3}$$

$$M\left[(H_2O)_n SO_4\right]_{ads} \longrightarrow M^{2+} + OH^- + SO_4^{2-} + H^+ \tag{4}$$

However, in presence of ionic liquids, anodic reactions can be represented as follows:

$$M + nH_2O \longleftrightarrow M(H_2O)n_{ads} \tag{5}$$

$$M(H_2O)n_{ads} + SO_4{}^{2-} \longleftrightarrow M\left[(H_2O)_n SO_4^{2-}\right]_{ads} \tag{6}$$

$$M(H_2O)_n SO_4^{2-}]_{ads} + ILsC^+ \longrightarrow M(H_2O)_n SO_4 ILsC]_{ads}^- \tag{7}$$

$$M(H_2O)_n SO_4 ILsC]_{ads}^- + ILsC^+ + SO_4^{2-} \longrightarrow (M(H_2O)_n SO_4 ILsC]^-{}_{ads} ILsC^+ SO_4^{2-}/ILsC^+ \tag{8}$$

$$M + X^- \longleftrightarrow (MX^-)_{ads} \tag{9}$$

$$(MX^-)_{ads} + ILsC^+ \longleftrightarrow \left(MX^- ILsC^+\right)_{ads} \tag{10}$$

where, $ILsC^+$ and X^- represent the cationic counter part of the ionic liquids (mostly organic) and anionic counter part of the ionic liquid, respectively. It is important to mention that the concentration of sulphate ions is much higher as comared to the concentration of anionic counter part of the ionic liquids (X^-) that results into formation of $[M(H_2O)\ SO^{2-}{}_4]_{ads}$ in larger proporsion than $[MX^-]_{ads}$. Nevertheless, these both anionic charged species attracted positively charged cationic counter part of the ionic liquids ($ILsC^+$) by electrostatic force of attraction (physisoprtion) and forms monomolecular layer as an insoluble complex on the metallic surface [78, 145]. The adsortion of the $ILsC^+$ on metallic surface causes change in the surface polarity which induces the adsorption of the sulphate and X^- ions again which results into multimolecular layer [78, 146]. The multimolecular layers are stabilized by Vanderwaal's cohesion force acting beteween organic moeity of the ionic liquids which causes a more closely adsorbed film at metal/electrolyte interfaces. Generally, the cationic part ($ILsC^+$) interacts with the metallic surface and forms the multimolecular layers while rest of the part of the ionic liquids form hydrophobic hemi-micelles, ad-micelles and/or surface aggregation [78, 147]. The adsorbed multimolecular layers of the ILs isolate the metal (M) from corrosive enviroment and protect from corrosive dissolution.

The cathodic hydrogen evolution reaction (HER) can be represented by following simple stoichimmetry equation [148]:

$$H_2O + 2e^- \longleftrightarrow H_{2(g)} + 2OH^- \tag{11}$$

Generally, the hydrogen evolution reaction (HER) follows two very common mechanisms that is, Volmer-Heyrovsky mechanism represented by Eqs. (12) and (13) or according to the Tafel hydrogen evolution mechanism represented by Eq. (14). In acidic medium, the Volmer-Heyrovsky and Volmer-Tafel hydrogen evolution mechanisms have been shown below [148–150]:

$$M + H_3O^+ + e^- \longleftrightarrow MH_{ads} + H_2O \quad (\text{Volmer, V}) \tag{12}$$

$$MH_{ads} + H_3O^+ + e^- \longleftrightarrow H_2 + M + H_2O \quad \left(\text{Heyrovsky, H}\right) \tag{13}$$

$$MH_{ads} + MH_{ads} \longleftrightarrow H_2 + 2M \quad (\text{Tafel, T}) \tag{14}$$

During the first step of cathodic reactions hydrogen ions (or hydronium ions) first adsorbed on the metallic surface by Volmer mechanism followed by discharge of hydrogen gas by Heyrovsky and Tafel mechanism represented by Eqs. (13-14). All these reactions do not occur with the same rate. Generally, a slow reaction step is followed by a fast reaction step [151]. If the Volmer reaction is fast, then Heyrovsky and/or Tafel reactions occur with slower rate and vice versa. Presence of the organic corrosion inhibitors (ILs) in the corrosive solution may retards or slow down the formation of MH_{ads} or retards the electron transfer to the hydronium ions and suppresses the Heyrovsky reactions (13). In general, in corrosive medium, the adsorbed hydrogen on metallic surface recombined and evolved as the bubbles of hydrogen gas. The formation of bubble and its evolution is the second step in the HER. The formation of hydrogen gas either occurs through hydrogen atom-atom combination as denoted by Volmer-Tafel Eq. (14) or may results through hydrogen atom-hydrogen ion combination as represented by Volmer-Heyrovsky Eq. (13) [151].

In the presence of inhibitors (ILs), cathodic can be represented as follows:

$$M + ILsC^+ + e^- \longleftrightarrow M(ILsC)_{ads} \tag{15}$$

Initially, adsorption of hydronium ions and evolution of hydrogen gas occur at cathodic sites, simultaneously. At cathode, the cationic part of ionic liquids ($ILsC^+$) starts competing with hydrogen ions for electrons [78, 152]. In general, $ILsC^+$ has large molecular size and therefore replaces greater number of water molecules from the metallic surface. After their adsorption, cationic part of the ILs accepts electrons from the metal (M) which results into the formation of electrically neutral ionic liquids (inhibitors). The neutral species transfer (donation) their non-bonding (of heteroatoms) and π-electrons into the d-orbitals of the surface metallic atoms resulting into the formation of co-ordinate bonds between metal and ILs (chemisorption) as reported for several organic conventional inhibitors [78, 146, 153–156]. However, metals are already electron rich species; this type of donation causes inter electronic repulsion which interns resulted into transfer of electrons from d-orbitals of the surface metallic atoms to antibonding molecular orbitals of the ILs (retro-donation). Both donation and retro-donation strengthen each other through synergism [153–160].

5. Conclusions and future perspectives

On the basis of ongoing discussion it can be concluded that ionic liquids are green and sustainable inhibitors for corrosion of metals and alloys. The superiority of the use of ionic liquids as corrosion inhibitors compared to traditional volatile (toxic) corrosion inhibitors is based on the fact that they possess several fascinating properties such lower volatility, non-inflammability, non-toxic nature, chemical stability, high solubility in the polar solvents and their ability to easily adsorb on the metallic surface. Adsorption of the ionic liquids over the metallic surface results into formation of protective film which isolates the metals (alloys) from the corrosive environment and thereby inhibits corrosion. Among several available ionic liquids, imidazole based ionic liquids have been most extensively used. Some reports described the adsorption behavior of ionic liquids on metallic surface using DFT based quantum chemical calculations. However, the use of this technique should be further explored owing to its green nature to understand the mechanistic aspects of corrosion inhibition. The use of ionic liquids as corrosion inhibitors is preferred comparing with traditional inhibitors due to several physiochemical properties advantageous including their high solubility, non-toxic, high conductivity, and non-flammability, less volatility as well as high chemical stability and more importantly due to their "green and sustainable" nature.

Author details

Chandrabhan Verma[1,2], Eno E. Ebenso[2] and Mumtaz Ahmad Quraishi[1,3]*

*Address all correspondence to: maquraishi.apc@itbhu.ac.in

1 Department of Chemistry, Indian Institute of Technology, Banaras Hindu University, Varanasi, India

2 Material Science Innovation and Modelling (MaSIM) Research Focus Area, Department of Chemistry, Faculty of Agriculture, Science and Technology, North-West University (Mafikeng Campus), Private Bag, Mmabatho, South Africa

3 Center of Research Excellence in Corrosion, Research Institute, King Fahd University of Petroleum and Minerals, Dhahran, Saudi Arabia

References

[1] Revie RW, Uhling HH. Corrosion and Corrosion Control. 4th ed. John Wiley & Sons, Inc., Hoboken New Jersey, Canada; 2007. pp. 1-3

[2] Masadeh S. The Effect of Added Carbon Black to Concrete Mix on Corrosion of Steel in Concrete, Journal of Minerals and Materials Characterization and Engineering. 2015;**3**: 271-276

[3] Winkleman A, Svedberg EB, Schafrik E, Duquette DJ. Advanced Materials & Processes. 2011;**2011**:26-31

[4] Panchenko YM, Marshakov AI. Corrosion Science. 2016;**109**:217-229

[5] NACE. Materials Performance. Special Issue, July, Houston, Texas, USA; 2002

[6] Confederation of Indian Industry, 1st Global Corrosion Summit, New Delhi, India; 2011

[7] http://insights.globalspec.com/article/2340/annual-global-cost-of-corrosion-2-5-trillion

[8] http://www.mintek.co.za/2011/11/15/the-high-cost-of-corrosion

[9] Bakkar A, Ataya S. Corrosion Science. 2014;**85**:343-351

[10] Pavithra MK, Venkatesha TV, Punith Kumar MK, Tondan HC. Corrosion Science. 2012;**60**:104-111

[11] Obot IB, Macdonalda DD, Gasem ZM. Corrosion Science. 2015;**99**:1-30

[12] Popova A, Christov M, Zwetanova A. Corrosion Science. 2007;**49**:2131-2143

[13] AlaouiMouayd A, Orazem ME, Sutter EMM, Tribollet B, Koltsov A. Corrosion Science. 2014;**82**:362-368

[14] Popova A, Christov M, Vasilev A. Corrosion Science. 2015;**94**:70-78

[15] Finšgar M, Jackson J. Corrosion Science. 2014;**86**:17-41

[16] Verma C, Quraishi MA, Ebenso EE, Obot IB, El Assyry A. Journal of Molecular Liquids. 2016;**219**:647-660

[17] Cuil Y, Liu S, Smith K, Hu H, Tang F, Li Y, Yu K. Journal of Environmental Sciences. 2016;**48**:79-91; DOI: 10.1016/j.jes.2015.12.035

[18] Mazères B, Desgranges C, Toffolon-Masclet C, Monceau D. Corrosion Science. 2016;**103**: 10-19

[19] Yu X, Jiang B, Yang H, Yang Q, Xia X, Pan F. Applied Surface Science. 2015;**353**:1013-1022

[20] Zakariaa K, Hamdya A, Abbas MA, Abo-Elenien OM. Journal of the Taiwan Institute of Chemical Engineers. 2016;**65**:530-543

[21] Duonga NT, Hanga TTX, Nicolay A, Paint Y, Olivier MG. Progress in Organic Coatings. 2016;**101**:331-341

[22] de Souza TM, Cordeiro RFB, Viana GM, Aguiar LCS, de Senna LF, Malta LFB, D'Elia E. Journal of Molecular Structure. 2016;**1125**:331-339

[23] El Haleem SMA, ElWanees SA, El EEAA, Farouk A, Corrosion Science. 2013;**68**:1-13

[24] Zaferani SH, Shishesaz MR. Journal of Petroleum and Environmental Biotechnology. 2014;**5**:1-5

[25] Park H, Kim KY, Choi W. The Journal of Physical Chemistry B. 2002;**106**:4775-4781

[26] Abiola OK, Oforka NC, Ebenso EE, Nwinuka NM. Anti-Corrosion Methods and Materials. 2007;**54**:219-224

[27] Ihara M, Nishihara H, Aramaki K. Corrosion Science. 1992;**33**:1267-1279

[28] Rajeev P, Surendranathan AO, Murthy CSN. Journal of Materials and Environmental Science. 2012;**3**:856-869

[29] Cotting F, Aoki IV. Surface and Coatings Technology. 2016;**303**:310-318

[30] Xhanari K, Finsgar M. Arabian Journal of Chemistry. 2016. DOI: 10.1016/j.arabjc.2016.08.009

[31] Hamani H, Douadi T, Daoud D, Al-Noaimi M, Chafaa S. Measurement. 2016;**94**:837-846

[32] de Souza FS, Snelli A. Corrosion Science. 2009;**51**:642-649

[33] Eddy NO, Ebenso EE. African Journal of Pure and Applied Chemistry. 2008;**2**:46-54

[34] Abiola OK, Otaigbe JOE, Kio OJ. Corrosion Science. 2009;**51**:1879-1888

[35] Domling A. Chemical Reviews. 2006;**106**:17-89

[36] Singh MS, Chowdhury S. RSC Advances. 2012;**2**:4547-4592

[37] Cioc RC, Ruijter E, Orru RVA. Green Chemistry. 2014;**16**:2958-2975

[38] Capello C, Fischer U, Hungerbuhler K. Green Chemistry. 2007;**9**:927-934

[39] Rajeswari V, Kesavanb D, Gopiramanc M, Viswanathamurthi P, Poonkuzhali K, Palvannan T. Applied Surface Science. 2014;**314**:537-545

[40] Hamdani NE, Fdil R, Tourabi M, Jamac C, Bentiss F. Applied Surface Science. 2015; **357**:1294-1305

[41] Anupama KK, Ramya K, Shainy KM, Joseph A. Materials Chemistry and Physics. 2015;**167**:28-41

[42] Guo L, Zhu S, Zhang S. Journal of Industrial and Engineering Chemistry. 2015;**24**:174-180

[43] Grassino AN, Halambek J, Djakovic S, Brncic SR, Dent M, Fa ZG. Food Hydrocolloids. 2016;**52**:265-274

[44] Hulsbosch J, Vos DED, Binnemans K, Ameloot R. ACS Sustainable Chemistry & Engineering. 2016;**4**:2917-2931

[45] Xing DY, Yi Dong W, Chung TS. Industrial & Engineering Chemistry Research. 2016; **55**:7505-7513

[46] Grieve TL, Drummond CJ, Chemical Reviews. 2015;**115**:11379-11448

[47] Drioli E, Macedonio F. Industrial & Engineering Chemistry Research. 2012;**51**(30):10051-10056

[48] Amarasekara AS, Chemical Reviews. 2016;**116**:6133-6183

[49] Tanga S, Liua S, Guoa Y, Liua X, Jiang S. Journal of Chromatography A. 2014;**1357**:147-157

[50] Jadhav AH, Lim AC, Thorat GM, Jadhava HS, Seo JG. RSC Advances. 2016;**6**:1675-31686

[51] Varma RS. ACS Sustainable Chemistry & Engineering. 2016;**4**:5866-5878

[52] Hart WES, Harper JB, Aldous L. Green Chemistry. 2015;**17**:214-218

[53] Cevascoa G, Chiappe C. Green Chemistry. 2014;**16**:2375-2385

[54] Gu Y, Jerome F. Chemical Society Reviews. 2013;**42**:9550-9570

[55] Price BK, Hudson JL, Tour JM. Journal of the American Chemical Society. 2005;**127**: 14867-14870

[56] Dewilde S, Dehaen W, Binnemans K. Green Chemistry. 2016;**18**:1639-1652

[57] Goossens K, Lava K, Bielawski CW, Binnemans K. Chemical Reviews. 2016;**116**:4643-4807

[58] Hayes R, Warr GG, Atkin R. Chemical Reviews. 2015;**115**:6357-6426

[59] Jordana A, Gathergood A. Chemical Society Reviews. 2015;**44**:8200-8237

[60] Sosnowska A, Barycki M, Zaborowska M, Rybinska A, Puzyn T. Green Chemistry. 2014;**16**:4749-4757

[61] Estager J, Holbrey JD, Swadźba-Kwaźny M. Chemical Society Reviews. 2014;**43**:847-886

[62] Suresh, Sandhu JS. Green Chemistry. Letters Review. 2011;**4**:289-310

[63] Hajipour AR, Refiee F. Organic Preparations and Procedures International. 2015;**47**: 249-308

[64] Winkler DA, Breedon M, White P, Hughes AE, Sapper ED, Cole I. Corrosion Science. 2016;**106**:229-235

[65] Umorena SA, Solomon MM. Journal of Environmental Chemical Engineering. 2017;**5**: 246-273

[66] Mobin M, Aslam R, Aslam J. Materials Chemistry and Physics. 2017;**191**:151-167

[67] Keskin S, Kayrak-Talay D, Akman U, Hortacsu O. The Journal of Supercritical Fluids. 2007;**43**:150-180

[68] Shi R, Wang Y. Scientific Reports. 2016;**6**:1-12

[69] Wan K, Feng P, Hou B, Li Y. RSC Advances. 2016;**6**:77515-77524

[70] Migahed MA, Attia AA, Habib RE. RSC Advances. 2015;**5**:57254-57262

[71] Verma CB, Quraishi MA, Ebenso EE. International Journal of Electrochemical Science. 2014;**9**:5507-5519

[72] Xhanari K, Finsgar M. Arabian Journal of Chemistry. 2016;**xxx**:xxx-xxx, DOI: 10.1016/j.arabjc.2016.08.009

[73] Lopes JNC, Gomes MFC, Padua AAH. The Journal of Physical Chemistry B. 2006;**110**: 16816-16818

[74] Verma C, Olasunkanmi LO, Ebenso EE, Quraishi MA, Obot IB. The Journal of Physical Chemistry B. 2016;**120**:11598-11611

[75] Verma C, Olasunkanmi LO, Obot IB, Ebenso EE, Quraishi MA. RSC Advances. 2016;**6**: 53933-53948

[76] Verma C, Olasunkanmi LO, Obot IB, Ebenso EE, Quraishi MA. RSC Advances. 2016;**6**: 15639-15654

[77] Verma C, Quraishi MA, Olasunkanmi LO, Ebenso EE. RSC Advances. 2015;**5**:85417-85430

[78] Likhanova NV, Domínguez-Aguilar MA, Olivares-Xometl O, Nava-Entzana N, Arce E, Dorantes H. Corrosion Science. 2010;**52**:2088-2097

[79] Tüken T, Demir F, Kıcır N, Sıgırcık G, Erbil M. Corrosion Science. 2012;**59**:110-118

[80] Zhang QB, Hua YX. Electrochimica Acta. 2009;**54**:1881-1887

[81] Zheng X, Zhang S, Li W, Gong M, Yin L. Corrosion Science. 2015;**95**:168-179

[82] El-Shamy AM, Zakaria K, Abbas MA, El Abedin SZ. Journal of Molecular Liquids. 2015;**211**:363-369

[83] Sasikumar Y, Adekunle AS, Olasunkanmi LO, Bahadur I, Baskar R, Kabanda MM, Obot IB, Ebenso EE. Journal of Molecular Liquids. 2015;**211**:105-118

[84] Zhoua X, Yanga H, Wang F. Electrochimica Acta. 2011;**56**:4268-4275

[85] Bousskri A, Anejjar A, Messali M, Salghi R, Benali O, Karzazi Y, Jodeh S, Zougagh M, Ebenso EE, Hammouti B. Journal of Molecular Liquids. 2015;**211**:1000-1008

[86] Tawfik SM. Journal of Molecular Liquids. 2016;**216**:624-635

[87] Kannan P, Karthikeyan J, Murugan P, SubbaRao T, Rajendran N. Journal of Molecular Liquids. 2016;**221**:368-380

[88] Javadiana S, Yousefia A, Neshati J. Applied Surface Science. 2013;**285**:674-681

[89] Kowsaria E, Payamia M, Aminib R, Ramezanzadehb B, Javanbakht M. Applied Surface Science. 2014;**289**:478-486

[90] Ullah S, Bustam MA, Shariff AM, Gonfa G, Izzat K. Applied Surface Science. 2016;**365**:76-83

[91] Ashassi-Sorkhabi H, Es'haghi M, Materials Chemistry and Physics. 2009;**114**;267-271

[92] Zuriaga-Monroy C, Oviedo-Roa R, Montiel-Sánchez LE, Vega-Paz A, Marín-Cruz J, Martínez-Magadán JM. Industrial & Engineering Chemistry Research. 2016;**55**:3506-3516

[93] Lozano I, Mazario E, Olivares-Xometl CO, Likhanova NV, Herrasti P. Materials Chemistry and Physics. 2014;**147**:191-197

[94] Yesudass S, Olasunkanmi LO, Bahadur I, Kabanda MM, Obot IB, Ebenso EE. Journal of the Taiwan Institute of Chemical Engineers. 2016;**64**:252-268

[95] Chong AL, Mardel JI, MacFarlane DR, Forsyth M, Somers AE. ACS Sustainable Chemistry & Engineering. 2016;**4**:1746-1755

[96] Ma Y, Han F, Li Z, Xia C. ACS Sustainable Chemistry & Engineering. 2016;**4**:5046-5052

[97] Ontiveros-Rosales M, Olivares-Xomet O, Likhanova NV, Lijanova IV, Guzman-Lucero D, Mendoza-Herrera MDC. Research on Chemical Intermediates. DOI: 10.1007/s11164-016-2643-3

[98] Acidi A, Hasib-ur-Rahman M, Larachi F, Abbaci A. Korean Journal of Chemical Engineering. 2014;**31**:1043-1048

[99] Yousefi A, Javadian S, Dalir N, Kakemam J, Akbari J. RSC Advances. 2015;**5**:11697-11713

[100] Yang D, Zhang M, Zhenga J, Castaneda H. RSC Advances. 2015;**5**:95160-95170

[101] Godec RF. Industrial & Engineering Chemistry Research. 2010;**49**:6407-6415

[102] Murulana LC, Singh AK, Shukla SK, Kabanda MM, Ebenso EE. Industrial & Engineering Chemistry Research. 2012;**51**:13282-13299

[103] Hasib-ur-Rahman M, Larachi F. Industrial & Engineering Chemistry Research. 2013;**52**: 17682-17685

[104] Yousefi A, Javadian S, Neshati J. Industrial & Engineering Chemistry Research. 2014; **53**:16349-16358

[105] Guzman-Lucero D, Olivares-Xomet O, Martínez-Palou R, Likhanova NV, Domínguez-Aguilar MA, Garibay-Febles V, Industrial & Engineering Chemistry Research. 2011;**50**: 7129-7140

[106] Olivares-Xomet O, López-Aguilar C, Herrastí-González P, Likhanova NV, Lijanova I, Martínez-Palou R, Antonio Rivera-Márquez J. Industrial & Engineering Chemistry Research. 2014;**53**:9534-9543

[107] Li J, Li D, Zhou F, Feng D, Xia Y, Liu W. Industrial Lubrication and Tribology. 2015;**67**: 210-215

[108] Mashuga ME, Olasunkanmi LO, Adekunle AS, Yesudass S, Kabanda MM, Ebenso EE. Materials. 2015;**8**:3607-3632

[109] El-Sayed MS, Abdo HS, Abedin SZE. Materials. 2015;**8**:3883-3895

[110] Velrani S, Muthukrishnan P, Jeyaprabha B, Prakash P. International Journal of Engineering Research and Technology. 2014;**3**:2704-2720

[111] Ibrahim MAM, Messali M, Moussa Z, Alzahrani AY, Alamry SN, Hammouti B. Portugaliae Electrochimica Acta. 2011;**29**:375-389

[112] Atta AM, El-Mahdy GA, Al-Lohedan HA, Ezzat ARO. Molecules. 2015;**20**:11131-11153

[113] Zarrouk A, Messali M, Zarrok H, Salghi R, Al-Sheikh Ali A, Hammouti B, Al-Deyab SS, Bentiss F. International Journal of Electrochemical Science. 2012;**7**:6998-7015

[114] Messali M, Asiri MAM. Journal of Materials and Environmental Science. 2013;**4**:770-785

[115] Ezhilarasi MR, Prabha B, Santhi T. Chemical Science Transactions. 2015;**4**:758-767

[116] El-Mahdy GA, Atta AM, Al-Lohedan HA, Ezzat AO. International Journal of Electrochemical Science. 2015;**10**:5812-5826

[117] Tseng CH, Chang JK, Chen JR, Tsai WT, Deng MJ, Sun IW. Electrochemistry Communication. 2010;**12**:1091-1094

[118] Lin PC, Sun IW, Chang JK, Su CJ, Lin JC. Corrosion Science. 2011;**53**:4318-4323

[119] Hanza AP, Naderi R, Kowsari E, Sayebanida M. Corrosion Science. 2016;**107**:96-106

[120] Kowsari E, Arman SY, Shahini MH, Zandi H, Ehsani A, Naderi R, Pourghasemi Hanza A, Mehdipour M. Corrosion Science. 2016;**112**:73-85

[121] Messaadia L, El Mouden OID, Anejjar A, Messali M, Salghi R, Benali O, Cherkaoui O, Lallam A. Journal of Materials and Environmental Science. 2015;**6**:598-606

[122] Likhanova NV, Olivares-Xometl O, Guzmán-Lucero D, Domínguez-Aguilar MA, Nava N, Corrales-Luna M, Mendoza MC. International Journal of Electrochemical Science. 2011;**6**:4514-4536

[123] Zhang QB, Hua YX. Materials Chemistry and Physics. 2010;**119**:57-64

[124] Shetty SK, Shetty AN. Canadian Chemical Transactions. 2015;**3**:41-64

[125] Arellanes-Lozada P, Olivares-Xometl O, Guzmán-Lucero D, Likhanova NV, Domínguez-Aguilar MA, Lijanova IV, Arce-Estrada E. Materials. 2014;**7**:5711-5734

[126] Zhang Q, Gao Z, Xu F, Zou X. Colloids and Surfaces A: Physicochemical and Engineering Aspects. 2011;**380**:191-200

[127] Trombetta F, de Souza RF, Oberson de Souza M, Borges CB, Panno NF, Martini EMA. Corrosion Science. 2011;**53**:51-58

[128] Li X, Deng S, Fu H. Corrosion Science. 2011;**53**:1529-1536

[129] Bermudez MD, Jimenez AE, Martınez-Nicolas G. Applied Surface Science. 2007;**253**: 7295-7302

[130] Tavakoli H, Shahrabi T, Hosseini MG. Materials Chemistry and Physics. 2008;**109**:281-288

[131] Elmorsi MA, Hassanein AM. Corrosion Science. 1999;**41**:2337-2352

[132] Zucchi F, Grassi V, Frignani A, Trabanelli G. Corrosion Science. 2004;**46**:2853-2865

[133] Yi-Xin ZQBH. Acta Physico-Chimica Sinica. 2011;**27**:655-663

[134] Gabler C, Tomastik C, Brenner J, Pisarova L, Doerra N, Allmaier G. Green Chemistry. 2011;**13**:2869-2877

[135] Manamela KM, Murulana LC, Kabanda MM, Ebenso EE. International Journal of Electrochemical Science. 2014;**9**:3029-3046

[136] Huanga P, Lathama JA, MacFarlane DR, Howletta PC, Forsyth M. Electrochimica Acta. 2013;**110**:501-510

[137] Suna J, Howlett PC, MacFarlane DR, Lina J, Forsyth M. Electrochimica Acta. 2008;**54**: 254-260

[138] Shukla SK, Murulana LC, Ebenso EE. International Journal of Electrochemical Science. 2012;**6**:4286-4295

[139] Scendo M, Uznanska J, International Journal of Corrosion. 2011;**2011**:1-13

[140] Rosales MO, Xometl OO, Likhanova NV, Lijanova IV, Lucero G, Mendoza-Herrer MDC. Research on Chemical Intermediates. 2017;**43**:641-660

[141] Badr EA. Journal of Industrial and Engineering Chemistry. 2014;**20**:3361-3366

[142] Peme T, Olasunkanmi LO, Bahadur I, Adekunle AS, Kabanda MM, Ebenso EE. Molecules. 2015;**20**:16004-16029

[143] Atta AM, El-Mahdy GA, Allohedan HA, Abdullah MMS. International Journal of Electrochemical Science. 2016;**11**:882-898

[144] Abd El-Maksoud SA. International Journal of Electrochemical Science. 2008;**3**:528-555

[145] Zhang D, Li L, Cao L, Yang N, Huang C. Corrosion Science. 2001;**43**:1627-1636

[146] Fuchs-Godec R. Colloids and Surfaces A. 2006;**280**:130-139

[147] Gennero de Chialvo MR, Chialvo AC. Electrochimica Acta. 1998;**44**:841-851

[148] de Chialvo MRG, Chialvo AC. Electrochemistry Communications. 1999;**1**:379-382

[149] Bhardwaj M, Balasubramaniam R. International Journal of Hydrogen Energy. 2008; **33**:2178-2188

[150] Atta NF, Fekry AM, Hassaneen HM. International Journal of Hydrogen Energy. 2011; **xxx**:1-10. DOI: 10.1016/j.ijhydene.2011.02.134

[151] Migahed MA. Materials Chemistry and Physics. 2005;**93**:48-57

[152] Antonijevic MM, Petrovic MB. International Journal of Electrochemical Science. 2008;**3**:1-28

[153] Verma C, Ebenso EE, Vishal Y, Quraishi MA. Journal of Molecular Liquids. 2016;**224**: 1282-1293

[154] Haque J, Srivastava V, Verma C, Quraishi MA. Journal of Molecular Liquids. 2017;**225**: 848-855

[155] Verma C, Quraishi MA, Singh A. Journal of the Taiwan Institute of Chemical Engineers. 2016;**58**:127-140

[156] Gupta NK, Verma C, Quraishi MA, Mukherjee AK. Journal of Molecular Liquids. 2016; **215**:47-57

[157] Nwankwo HU, Ateba CN, Olasunkanmi LO, Adekunle AS, Isabirye DA, Onwudiwe DC, Ebenso EE. Materials. 2016;**9**:1-19

[158] Karthik G, Sundaravadivelu M. Egyptian Journal of Petroleum. 2016;**25**:183-191

[159] Thirumalairaj B, Jaganathan M. Egyptian Journal of Petroleum. 2016:**25**:423-432

[160] Verma C, Singh A, Pallikonda G, Chakravarty M, Quraishi MA, Bahadur I, Ebenso EE. Journal of Molecular Liquids. 2015;**209**:306-319

Section 5

Pollution Prevention Resource

Chapter 7

Nanoscale Zero Valent Iron for Environmental Cadmium Metal Treatment

Keyla T. Soto-Hidalgo and Carlos R. Cabrera

Additional information is available at the end of the chapter

http://dx.doi.org/10.5772/intechopen.72737

Abstract

In the course of developing methods to treat heavy metal contaminants in wastewater, nanoscale zerovalent iron (nZVI) has been found to be an alternative approach. This nanoparticle has been used to remove metals such as Cr^{6+}, Cu^{2+}, Pb^{2+}, Ba^{2+}, As^{3+}, As^{5+}, and Co^{2+} from aqueous solutions. Iron nanoparticles are useful for decontamination purposes due to their smaller size, surface area-to-weight ratio, and capacity to remove groundwater contaminants. The large specific surface area of the iron nanoparticles further fosters enhanced reactivity for the transformation of environmental pollutants. Because of their smaller size, nanoscale-based iron materials are much more reactive than conventional iron powders, and they can be suspended in slurry and pumped straight to the contaminated site. The ZVI is often applied for the remediation of wastewater or groundwater with several kinds of reducible contaminants, which are near its surface reduction potential. This chapter seeks to present the efficiency of zerovalent iron nanoparticles (nZVI) to remedy the cadmium ion pollution in water as well as the use of the remediation product in photoelectrochemical devices.

Keywords: nanoscale zero valent iron, heavy metals, environmental remediation, photoelectrochemical solar cells

1. Introduction

Environmental pollution is one of the most important problems in the world and is the focus of a wide array of studies in the scientific community [1]. The development of advanced technology and rapid industrialization are the most predominant factors that increase environmental pollution [2]. One of the major hazards to human health from environmental contamination is heavy metals due to their tendency to bioaccumulate in plants and animals that are part of the human food chain [3]. Numerous anthropogenic activities such as mining, landfills, electroplating, metal processing, textile, petroleum refining, pesticides, battery and paint manufacturing, and printing and photographic industries release these metals into the environment. Heavy metals can persist for a long time in the environment [4].

Many metals that do not play any physiological role such as lead (Pb), cadmium (Cd), arsenic (As), and mercury (Hg) have adverse effects on human health because they have been cataloged as risk factors for the development of cancer, respiratory conditions, neurodegenerative disorders, and arthritis [5]. Lead is the most common of these toxic materials, and all Pb species are generally toxic (see **Table 1**). Volcanic activity and geochemical weathering are the greatest natural sources; man-made sources include lead-based paints, gasoline additives, food-can soldering, and battery making [6]. The movement of Pb from absorbing root hairs, is apparently impeded by several biochemical and/or physical processes involving Pb binding, inactivation, and/or precipitation [7]. Lead has accumulated in different terrestrial and aquatic ecosystems, and has been shown to accumulate in plants from several sources, including soil; however, the reports on accumulation of Pb within plants are variable [8, 9].

Arsenic is highly toxic to human health (see **Table 1**) [8, 10]. All species of As (III) and As (V) are toxic. Inorganic and most toxic forms of arsenic (arsenate and arsenite) are found in soil, crops, and water, particularly in groundwater from deep wells, often used as drinking water. These compounds are also found in environmental tobacco smoke and arsenic-treated wood, used in most outdoor wooden structures in the United States. High levels of arsenic are present in agricultural fertilizer that is used for soil treatment; as a consequence, any vegetables and fruits, if grown in this soil, will contain high levels of arsenic.

Cadmium ions, commonly found in soil and water systems, affect vital organs such as liver and kidneys (see **Table 1**) [8, 11]. This metal is considered one of the most toxic environmental substances due to its ubiquity, toxicity, and long half-life. All species of Cd are toxic. Exposure to cadmium occurs through inhalation (particularly in active cigarette smokers), groundwater consumption, industrial exposure, and contaminated food. It causes a wide variety of toxic effects when taken up by plants such as the inhibition of several plant physiological processes like oxidative reactions and nitrogen metabolism [12, 13]. Currently, there are many traditional chemical methods to remove these heavy metals from contaminated sites such as alkaline precipitation,

Pollutants	Major sources	Effect on human health	Permissible level (mg/l)
Arsenic	Pesticides, fungicides, metal smelters	Bronchitis, dermatitis, poisoning	0.02
Cadmium	Welding, electroplating, pesticide, fertilizer, Cd, and Ni batteries, and nuclear fission plant	Renal dysfunction, lung disease, lung cancer, bone defects, increased blood pressure, kidney damage, bronchitis, gastrointestinal disorder, bone marrow, and cancer	0.06
Lead	Paint, pesticide, smoking, automobile, emission, mining, and burning of coal	Mental retardation in children, developmental delay, fatal infant encephalopathy, congenital paralysis, damage to the nervous system, liver, and kidney	0.1

Table 1. Major sources of As, Cd, and Pb and their effect on human health [8].

ion exchange columns, electrochemical removal, filtration, and membrane technologies [8], but these methods are expensive and use many equipments to efficiently remove the contaminants. On the other hand, following the principles of Green Chemistry is necessary to use alternative products that prevent waste after remediation process, use less hazardous chemical synthesis, and minimize energy requirements of all chemical processes and environmental and economic impacts.

2. Iron nanomaterials for remediation process

Iron nanoparticles are a new generation of materials for environmental remediation. Various metallic ions, including Pb^{2+}, Cr^{6+}, Ni^{2+}, As^{3+}, As^{5+}, Cd^{2+}, Cu^{2+}, Zn^{2+}, and Ba^{2+} have been fixated from water using this new technology [14]. In situ remediation strategies are useful to reduce the mobile fraction of metals and metalloids in the soil that could reach the groundwater or be taken up by soil organisms. As such, several strategies have been used to promote the immobilization of metals in soil [15]. In the course of developing suitable options to remove heavy metal contaminants from wastewater, nanoscale zero valent iron (nZVI) particles have been found to be an alternative approach to reduce the concentration of several kinds of contaminants, mainly targeting chlorinated organic contaminants, inorganic anions, metals, and metalloids [15–19]. Although the benefits of this strategy are evident, governments and environmental agencies must evaluate any associated environmental risks because currently available ecotoxicology data are not enough [20].

Previous column experiments have showed the effectiveness of nZVI for the in situ immobilization of heavy metals, which reduces their potential leachability, as a strategy to prevent their transport into deeper soil layers, rivers, and groundwater [21]. Iron nanoparticles are particularly attractive for environmental remediation because these are much more reactive than iron powders and they can be suspended in slurry and moved to the polluted site [22, 23]. Recently, the synthesis and utilization of iron-based nanomaterials with novel properties and functions have been widely studied, both for their nanosize and for their magnetic characteristics [14]. In the environment, iron oxides are present naturally, but can also be chemically produced in nanoparticles of the order of 100 nm or less, which provide them with specific and better affinity for ions metals adsorption. For this reason, these nanoparticles are being used for in situ experiments [15, 16]. In environmental engineering, the application of nZVI is commonly used for the removal of metal/metalloids from polluted waters and soils, or their stabilization [17]. For example, during the remediation of contaminated soils, nZVI has become a widespread amendment for in situ applications, since it can form a permeable barrier in the soil in order to prevent the dissemination of contaminants by the soil pore water, thus achieving their immobilization [18, 24].

Efficient nZVI remediation of groundwater contaminants has been shown in multiple studies; however, regarding nZVI-induced soil toxicity, limited data have been reported, providing preliminary results about the effects of nZVI on soil biota and some plant species [19]. Most of the reported studies have been conducted either under no real conditions or only considering short-term exposure. Therefore, the impact of nZVI treatment on soil properties and functionality remains unclear. On the other hand, very few investigations of nZVI materials present a detailed study of the products formed in the remediation process for reusability of these nZVI after treatment [25–27]. An alternative use of the remediation product of nZVI could be for photoelectrochemical solar cells (PSC) applications.

In the nZVI reaction, metallic iron is oxidized in the presence of water, which can remove other metal ions from aqueous media by chemical absorption. In this process, iron oxides such as hematite (α-Fe_2O_3) are produced [27]. Hematite has been studied for catalytic applications because of the presence of active photochemical properties [28]. Even so, the application of hematite in PSC is a challenge for the scientist community because this species is highly active in UV range but not absorbed in the visible range. The aim is to employ α-Fe_2O_3 in commercial solar-based devices using high temperature synthesis methods for doping structures and an alternative method to produce dye-sensitized solar cells [28–31].

3. Nanoscale zero valent iron (nZVI) for cadmium decontamination process

Of all the metallic contaminants, cadmium draws special attention because of its high affinity and water solubility [32]. Cadmium species have been detected in aquatic ecosystems and found to bioaccumulate in organisms in nanomolar to micromolar concentrations [33]. Efficient nZVI remediations of groundwater contaminants have been shown in multiple studies [33–37]. However, in the literature, there is a lack of comparable studies for different nZVI materials and deployment strategies [38].

Various adsorption and kinetic models to describe metal adsorption on nZVI and Cd-nZVI surfaces using SEM/EDX and XPS measurements have been studied [26, 38] but, to our knowledge, the interactions between surface Fe^0 and other heavy metals in particular Cd^{2+} and subsequent cadmium retention in nZVI particles have not been the subject of detailed study.

Results from our study provide important information of the products formed during the remediation process of cadmium. Studies of redox and adsorption processes after treatment of nZVI have been evaluated [39]. However, it is necessary to understand in detail what occurs in the cadmium adsorption process at the nanoparticle surface. The reduction potential of Cd is larger than standard reduction potential (E^o) of nZVI (−0.40 V, 25°C, −0.447 V), respectively [26]. Our data suggest that Cd^{2+} ions are sequestrated on nZVI by adsorption process [27].

Results of the Cd concentration reduction on nZVI sample were analyzed using an inductively coupled plasma (ICP). The maximum cadmium adsorption percentage relative to the initial Cd^{2+} concentration of 6 ppm was 93%. This percentage was obtained after a period of 5 h, which indicates that longer interaction times between cadmium ions and nZVI promoted larger cadmium concentration reduction.

These results show that nZVI is an alternative to decrease high Cd concentration in contaminated sites. However, there are no sufficient data about the possible formation of toxic product after treatment with nZVI. For this reason, a structural analysis of used nZVI was deemed imperative to gain an understanding of the interactions between the nZVI and cadmium. This new knowledge may serve to optimize the remediation process and to provide alternative uses for the remediation product. The formations of unexpected nanofibers and cadmium ferrite structures have been reported. This remediation product or environmental waste has been suggested as a photocatalyst material that has great potential application for light harvesting [40]. These results could be useful because we can prevent the waste formation after chemical process, and reuse the products of remediation processes for other energy applications. This will decrease the amount of new hazardous substances produced after water decontamination processes.

A structural model of Cd-nZVI fibers is illustrated in **Figure 1**. This conceptual model of Fe^0 nanofibers was presented considering the results of X-ray diffraction patterns, X-photoelectron spectroscopy results, X-ray absorption spectroscopy, and high-resolution transmission electron microscopy (HRTEM) images. **Figure 1** shows the (A) Fe^0 nanoparticles, (B) the Fe^0–H_2O, where we observe Fe_2O_3 and FeOOH surrounding small quantities of Fe^0, and (C) Fe^0–Cd where we observe $CdFe_2O_4$, FeOOH, Fe_2O_3, and Fe^0. For the formation of the structure presented in **Figure 1C**, cadmium ions act as the limiting reagent, where the onset of the reaction of Cd, Fe^0, Fe_2O_3, and FeOOH allowing the possible formation of cadmium ferrite.

The oxyhydroxide iron (FeOOH) has a crystal structure containing tunnel-shaped cavities that run parallel to the c-axis. These sites are bound by double rows of fused octahedral, in which cadmium ions probably reside [41]. In **Figure 2a**, HRTEM images of nZVI exhibit spherical shapes and well-aligned aggregates with a diameter range between 25 and 70 nm.

These clusters of nanoparticles are caused by magnetic dipole-dipole interactions of the individual particles [42]. After nZVI Cd^{2+} exposure, shown in **Figure 2b**, nanofibers are organized [43]. These nanofibers were possibly produced by the diffusion of absorbed Cd^{2+} ions through the core-shell structure [36]. In iron oxide, an electron transfer reactions between Cd^{2+} ions and the Fe^0 might probably occur in the core (**Figure 2**).

The fiber formation as a product of nZVI in the presence of cadmium ions is possibly due to the rearrangement of the nanomaterial structure as a consequence of the adsorption process. The interactions of Cd^{2+} and Fe^{3+}, particularly, possibly promote the formation of $CdFe_2O_4$,

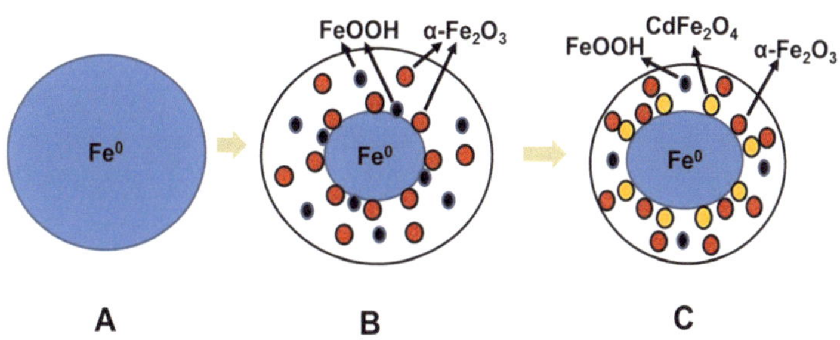

Figure 1. Conceptual model of cadmium adsorption process on nZVI nanostructures with cadmium-iron oxides on the surface [43].

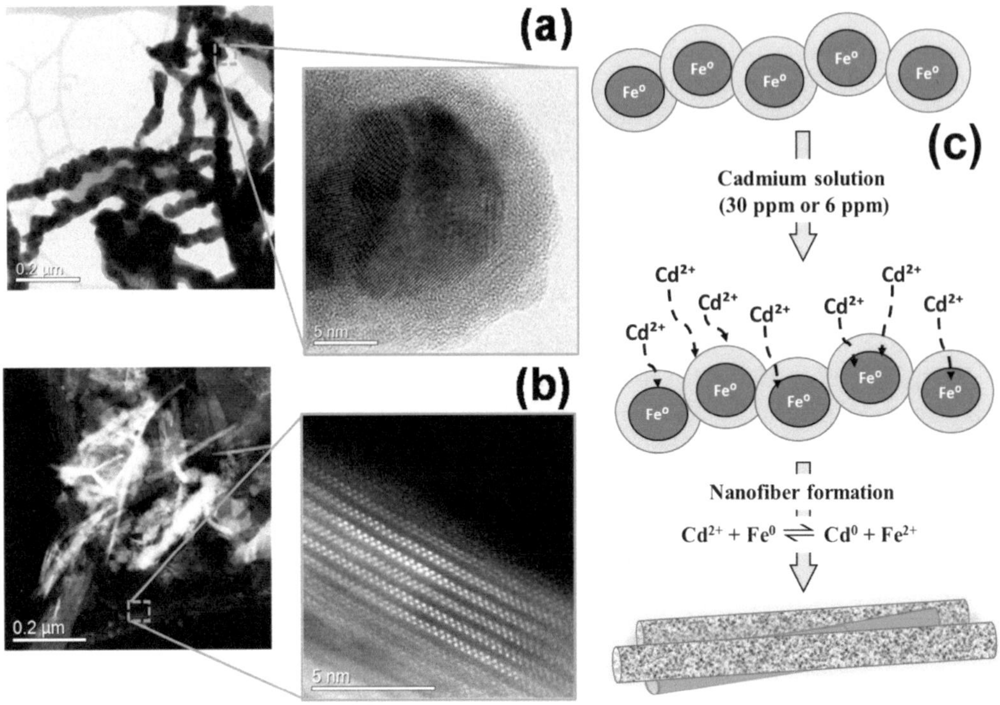

Figure 2. HRTEM images of (a) nZVI particles synthesized and (b) iron nanofibers formed during the cadmium ion remediation process. (c) Conceptual model representing the Cd^{2+} atoms interaction in Fe^0 core of nZVI [43].

which is in accordance with what is reported in the literature; these results show fiber formation analogous to previous studies with $CdFe_2O_4$ particles synthesized by a coprecipitation method [41]. These studies in HRTEM presented the aggregation of fine particles of $CdFe_2O_4$ having two kinds of shapes, fibrous and granular. The unintended formation of $CdFe_2O_4$ nanofibers as a remediation product presents an opportunity to reuse the remediation products for applications pertaining to light harvesting.

4. Applications of nZVI in photoelectrochemical solar cell devices

Photoelectrochemical solar cells use light to carry out a chemical reaction, converting light to chemical energy or power [44–46]. A photoelectrochemical cell is a photocurrent-generating device that has a semiconductor in contact with an electrolyte. It consists of a photoactive semiconductor working electrode (either n-type or p-type) and counter electrode made of either metal (e.g., Pt or C) or semiconductor. These electrodes are immersed in the electrolyte containing redox species with its standard potential being within the semiconductor bandgap potential region. In a metal-electrolyte junction, the potential drop occurs entirely on the solution site, whereas in a semiconductor-electrolyte junction, the potential drop occurs on the semiconductor site as well as the solution site [47].

The charge on the semiconductor side is distributed in the interior of the semiconductor, creating a space charge region. If the junction of the semiconductor-electrolyte is illuminated with a light having energy greater than the semiconductor bandgap, photogenerated electron-hole pairs are separated in the space charge region [48, 49]. The photogenerated minority carriers arrive at the interface of the semiconductor-electrolyte where a redox reaction will occur.

Photoelectrochemical cells, such as those produced by Brian O'Regan and Michael Grätzel, have been of interest to scientist because of their low manufacturing cost [44]. Photoelectrochemical devices require exhaustive optimization of their quantum conversion efficiency, which is affected by the electron transfer processes.

In the nanoscale iron reaction with water, metallic iron is oxidized to α-Fe_2O_3 [27]. Using α-Fe_2O_3 as an alternative of TiO_2 has been evaluated to produce dye-sensitized solar cells [31]. Methods such as doping with other metals and changing the structural arrangement of the system have been employed to overcome challenges regarding electron transfer processes [28]. The incorporation of Cd ions on the surface of the oxidized nZVI may produce surface structure changes. It has been found that oxide structures such as Fe_xO_y and the formation of $CdFe_2O_4$ may be present at the surface of the nanoparticles [43]. As described in recent reports, this surface process may occur without using high-temperature processes, a common surface reaction described in the literature [29, 30]. Moreover, photovoltaic and photoelectrochemical processes have been studied with $CdFe_2O_4$. However, few reports have shown the use of heavy metal doped ferrite particles as semiconductors in photovoltaic and photoelectrochemical devices [30, 50–52]. Recently, the use of Cd^{2+} ions exposed nZVI as semiconductors in PSCs has been reported. Brian O'Regan and Michael Grätzel reported similar systems in the 1990s, which caught the attention of the scientific community due to their low cost of fabrication [53].

Photoelectrochemical devices are challenging due to the optimization of their quantum conversion efficiency, which is affected by the electron transfer processes in the system. In our study, nZVI was exposed to different Cd^{2+} concentrations (1–30 ppm), similar to values found in contaminated areas of Puerto Rico. The novelty of these results was to analyze the material produced after the Cd decontamination processes in water using nZVI as a photoactive substance. The product formed exhibited capable photoactive behavior for photoelectrochemical solar cell applications.

In **Figure 3**, incident to photocurrent efficiency (IPCE) normalized signals of two PSCs are observed, each one with different material in the photoanode [54]. The samples prepared using the nZVI do not display significant signals (lowest curve). Particles treated with 30 ppm Cd^{2+} solution, however, exhibit a relative broadband from approximately 300–450 nm (black curve). This region is similar to the absorption results obtained in the UV/Vis analysis. Such a high photovoltage can be explained by an improvement in the electron transfer dynamics of the material in the PSC at higher cadmium concentrations due to structural changes as previously suggested in the literature [30]. As one of the principle of Green Chemistry, these results provide a new alternative to reuse nanomaterials used in decontamination processes and generate modified iron oxide photocatalyst without using high temperature. These data have a significant value for future applications in photoactive materials synthesis.

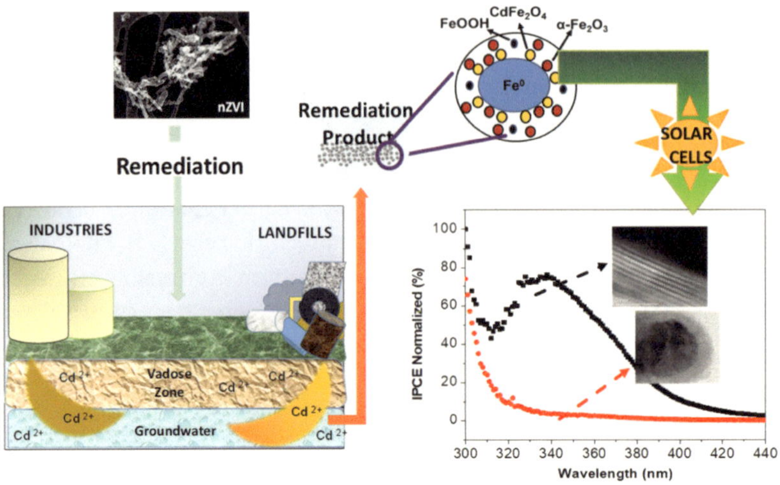

Figure 3. Graphical representation of cadmium water remediation to photoelectrochemical solar cells using nanoscale zero valent iron [54].

Acknowledgements

This work was supported by NSF-CREST Center for Innovation Research and Education in Environmental Nanotechnology Grant Number 1736093.

Author details

Keyla T. Soto-Hidalgo[1]* and Carlos R. Cabrera[2]

*Address all correspondence to: keyla.soto@upr.edu

1 NSF-CREST Center for Innovation, Research and Education in Environmental Nanotechnology (CIRE2N), University High School, University of Puerto Rico, San Juan, Puerto Rico

2 Department of Chemistry, University of Puerto Rico, San Juan, Puerto Rico

References

[1] Briggs D. Environmental pollution and the global burden of disease. British Medical Bulletin. 2003;**68**(1):1-24

[2] Kaznina NM, Laidinen GF, Titov AF, Talanov AV. Effect of lead on the photosynthetic apparatus of annual grasses. The Biological Bulletin. 2005;**32**(2):147-150

[3] Aycicek M, Kaplan O, Yaman M. Effect of cadmium on germination, seedling growth and metal contents of sunflower (*Helianthus annuus* L.). Asian Journal of Chemistry. 2013;**20**(4):2663-2672

[4] Wuana RA, Okieimen FE. Heavy metals in contaminated soils: A review of sources, chemistry, risks and best available strategies for remediation. ISRN Ecology. 2011;**2011**:20

[5] Barbosa JS, Cabral TM, Ferreira DN, Agnez-Lima LF, Batistuzzo de Medeiros SR. Genotoxicity assessment in aquatic environment impacted by the presence of heavy metals. Ecotoxicology and Environmental Safety. 2010;**73**(3):320-325

[6] Verma S, Dubey RS. Lead toxicity induces lipid peroxidation and alters the activities of antioxidant enzymes in growing rice plants. Plant Science. 2003;**164**(4):645-655

[7] Shen YT, Song YF. Effects of organic ligands on Pb absorption and speciation changes in Arabidopsis as determined by micro X-ray fluorescence and X-ray absorption near-edge structure analysis. Journal of Synchrotron Radiation. 2017;**24**(2):463-468

[8] Singh R, Gautam N, Mishra A, Gupta R. Heavy metals and living systems: An overview. Indian Journal of Pharmacology. 2011;**43**(3):246-253

[9] Maldonado-Magaña A, Favela-Torres E, Rivera-Cabrera F, Volke-Sepulveda TL. Lead bioaccumulation in *Acacia farnesiana* and its effect on lipid peroxidation and glutathione production. Plant and Soil. 2011;**339**(1):377-389

[10] Agency for Toxic Substances and Disease Registry (ASTDR), U. S. D. O. H. A. H. S., Toxicological profile for As. http://www.atsdr.cdc.gov/toxprofiles/tp2.pdf, Ed. 2007

[11] Weis JS, Weis P. Metal uptake, transport and release by wetland plants: Implications for phytoremediation and restoration. Environment International. 2004;**30**(5):685-700

[12] Laspina NV, Groppa MD, Tomaro ML, Benavides MP. Nitric oxide protects sunflower leaves against Cd-induced oxidative stress. Plant Science. 2005;**169**(2):323-330

[13] Hegedüs A, Erdei S, Horváth G. Comparative studies of H_2O_2 detoxifying enzymes in green and greening barley seedlings under cadmium stress. Plant Science. 2001;**160**(6):1085-1093

[14] Uzum C, Shahwan T, Eroglu AE, Lieberwirth I, Scott TB, Hallam KR. Application of zero-valent iron nanoparticles for the removal of aqueous Co^{2+} ions under various experimental conditions. Chemical Engineering Journal. 2008;**144**(2):213-220

[15] Yang X, Feng Y, He Z, Stoffella PJ. Molecular mechanisms of heavy metal hyperaccumulation and phytoremediation. Journal of Trace Elements in Medicine and Biology. 2005;**18**(4):339-353

[16] Kumpiene J, Lagerkvist A, Maurice C. Stabilization of As, Cr, Cu, Pb and Zn in soil using amendments – A review. Waste Management. 2008;**28**(1):215-225

[17] Mueller NC, Braun J, Bruns J, Černík M, Rissing P, Rickerby D, Nowack B. Application of nanoscale zero valent iron (NZVI) for groundwater remediation in Europe. Environmental Science and Pollution Research. 2012;**19**(2):550-558

[18] El-Temsah YS, Oughton DH, Joner EJ. Effects of nano-sized zero-valent iron on DDT degradation and residual toxicity in soil: A column experiment. Plant and Soil. 2013;**368**(1/2):189-200

[19] Fajardo C, Ortíz LT, Rodríguez-Membibre ML, Nande M, Lobo MC, Martin M. Assessing the impact of zero-valent iron (ZVI) nanotechnology on soil microbial structure and functionality: A molecular approach. Chemosphere. 2012;**86**(8):802-808

[20] Gil-Díaz M, Ortiz LT, Costa G, Alonso J, Rodríguez-Membibre ML, Sánchez-Fortún S, Pérez-Sanz A, Martín M, Lobo MC. Immobilization and leaching of Pb and Zn in an acidic soil treated with zerovalent iron nanoparticles (nZVI): Physicochemical and toxicological analysis of leachates. Water, Air, and Soil Pollution. 2014;**225**(6):1990

[21] Handy RD, van den Brink N, Chappell M, Mühling M, Behra R, Dušinská M, Simpson P, Ahtiainen J, Jha AN, Seiter J, Bednar A, Kennedy A, Fernandes TF, Riediker M. Practical considerations for conducting ecotoxicity test methods with manufactured nanomaterials: What have we learnt so far? Ecotoxicology. 2012;**21**(4):933-972

[22] Gil-Díaz M, Alonso J, Rodríguez-Valdés E, Pinilla P, Lobo MC. Reducing the mobility of arsenic in brownfield soil using stabilised zero-valent iron nanoparticles. Journal of Environmental Science and Health, Part A. 2014;**49**(12):1361-1369

[23] Matlochová A, Plachá D, Rapantová N. The application of nanoscale materials in groundwater remediation. Polish Journal of Environmental Studies. 2013;**22**(5):1401-1410

[24] El-Temsah YS, Joner EJ. Effects of nano-sized zero-valent iron (nZVI) on DDT degradation in soil and its toxicity to collembola and ostracods. Chemosphere. 2013;**92**(1):131-137

[25] Khin MM, Nair AS, Babu VJ, Murugan R, Ramakrishna S. A review on nanomaterials for environmental remediation. Energy & Environmental Science. 2012;**5**(8):8075-8109

[26] Boparai HK, Joseph M, O'Carroll DM. Kinetics and thermodynamics of cadmium ion removal by adsorption onto nano zerovalent iron particles. Journal of Hazardous Materials. 2011;**186**(1):458-465

[27] Li X-q, Zhang W-X. Sequestration of metal cations with Zerovalent iron Nanoparticles – A study with high resolution X-ray photoelectron spectroscopy (HR-XPS). Journal of Physical Chemistry C. 2007;**111**(19):6939-6946

[28] Lin Y, Xu Y, Mayer MT, Simpson ZI, McMahon G, Zhou S, Wang D. Growth of p-type hematite by atomic layer deposition and its utilization for improved solar water splitting. Journal of the American Chemical Society. 2012;**134**(12):5508-5511

[29] Ling Y, Wang G, Wheeler DA, Zhang JZ, Li Y. Sn-doped hematite nanostructures for Photoelectrochemical water splitting. Nano Letters. 2011;**11**(5):2119-2125

[30] Sivula K, Zboril R, Le Formal F, Robert R, Weidenkaff A, Tucek J, Frydrych J, Grätzel M. Photoelectrochemical water splitting with mesoporous hematite prepared by a solution-based colloidal approach. Journal of the American Chemical Society. 2010;**132**(21): 7436-7444

[31] Ocakoglu K, Krupnik T, van den Bosch B, Harputlu E, Gullo MP, Olmos JDJ, Yildirimcan S, Gupta RK, Yakuphanoglu F, Barbieri A, Reek JNH, Kargul J. Photosystem I-based biophotovoltaics on nanostructured hematite. Advanced Functional Materials. 2014;**24**(47):7467-7477

[32] Clemens S. Toxic metal accumulation, responses to exposure and mechanisms of tolerance in plants. Biochimie. 2006;**88**(11):1707-1719

[33] Crane RA, Scott TB. Nanoscale zero-valent iron: Future prospects for an emerging water treatment technology. Journal of Hazardous Materials. 2012;**211**(Supplement C):112-125

[34] Ponder SM, Darab JG, Mallouk TE. Remediation of Cr(VI) and Pb(II) aqueous solutions using supported, nanoscale zero-valent iron. Environmental Science & Technology. 2000;**34**(12):2564-2569

[35] Sun Y-P, Li X-q, Cao J, Zhang W-x, Wang HP. Characterization of zero-valent iron nanoparticles. Advances in Colloid and Interface Science. 2006;**120**(1):47-56

[36] Bishop EJ, Fowler DE, Skluzacek JM, Seibel E, Mallouk TE. Anionic homopolymers efficiently target zerovalent iron particles to hydrophobic contaminants in sand columns. Environmental Science & Technology. 2010;**44**(23):9069-9074

[37] Schrick B, Hydutsky BW, Blough JL, Mallouk TE. Delivery vehicles for zerovalent metal nanoparticles in soil and groundwater. Chemistry of Materials. 2004;**16**(11):2187-2193

[38] Boparai HK, Joseph M, O'Carroll DM. Cadmium (Cd^{2+}) removal by nano zerovalent iron: Surface analysis, effects of solution chemistry and surface complexation modeling. Environmental Science and Pollution Research. 2013;**20**(9):6210-6221

[39] Huff J, Lunn RM, Waalkes MP, Tomatis L, Infante PF. Cadmium-induced cancers in animals and in humans. International Journal of Occupational and Environmental Health. 2007;**13**(2):202-212

[40] Wu X, Yin S, Dong Q, Sato T. Preparation and visible light induced photocatalytic activity of C-$NaTaO_3$ and C-$NaTaO_3$-Cl-TiO_2 composite. Physical Chemistry Chemical Physics. 2013;**15**(47):20633-20640

[41] Yokoyama M, Sato T, Ohta E, Sato T. Magnetization of cadmium ferrite prepared by coprecipitation. Journal of Applied Physics. 1996;**80**(2):1015-1019

[42] Liu HB, Chen TH, Chang DY, Chen D, Liu Y, He HP, Yuan P, Frost R. Nitrate reduction over nanoscale zero-valent iron prepared by hydrogen reduction of goethite. Materials Chemistry and Physics. 2012;**133**(1):205-211

[43] Soto Hidalgo KT, Guzmán-Blas R, Ortiz-Quiles EO, Fachini ER, Corchado-García J, Larios E, Zayas B, José-Yacamán M, Cabrera CR. Highly organized nanofiber formation from zero valent iron nanoparticles after cadmium water remediation. RSC Advances. 2015;**5**(4):2777-2784

[44] Voloshin RA, Kreslavski VD, Zharmukhamedov SK, Bedbenov VS, Ramakrishna S, Allakhverdiev SI. Photoelectrochemical cells based on photosynthetic systems: A review. Biofuel Research Journal. 2015;**2**(2):227-235

[45] Lewis NS, Nocera DG. Powering the planet: Chemical challenges in solar energy utilization. Proceedings of the National Academy of Sciences of the United States of America. 2006;**103**(43):15729-15735

[46] Kamat PV. Semiconductor surface chemistry as holy grail in photocatalysis and photovoltaics. Accounts of Chemical Research. 2017;**50**(3):527-531

[47] Wei D, Amaratunga G. Photoelectrochemical cell and its applications in optoelectronics. International Journal of Electrochemical Science. 2007;**2**:897-912

[48] Sivula K, van de Krol R. Semiconducting materials for photoelectrochemical energy conversion. Nature Reviews Materials. 2016;**1**:15010

[49] van de Krol R. Photoelectrochemical measurements. In: van de Krol R, Grätzel M, editors. Photoelectrochemical Hydrogen Production. Boston: Springer; 2012. pp. 69-117

[50] Harish KN, Bhojya Naik HS, Prashanth kumar PN, Viswanath R. Optical and photocatalytic properties of solar light active Nd-substituted Ni ferrite catalysts: For environmental protection. ACS Sustainable Chemistry & Engineering. 2013;**1**(9):1143-1153

[51] Harish KN, Bhojya Naik HS, Prashanth kumar PN, Viswanath R. Synthesis, enhanced optical and photocatalytic study of Cd-Zn ferrites under sunlight. Catalysis Science & Technology. 2012;**2**(5):1033-1039

[52] Harish K, Bhojya Naik H, Prashanth Kumar P, Vishwanath R, Yashvanth Kumar G. Optical and photocatalytic properties of $CdFe_2O_4$ nanocatalysts: Potential application in water treatment under solar light irradiation. Archives of Applied Science Research. 2013;**5**(2):42-51

[53] O'Regan B, Gratzel M. A low-cost, high-efficiency solar cell based on dye-sensitized colloidal TiO_2 films. Nature. 1991;**353**(6346):737-740

[54] Soto Hidalgo KT, Ortiz-Quiles EO, Betancourt LE, Larios E, José-Yacaman M, Cabrera CR. Photoelectrochemical solar cells prepared from nanoscale zerovalent iron used for aqueous Cd^{2+} removal. ACS Sustainable Chemistry & Engineering. 2016;**4**(3):738-745

Chapter 8

The Utility of the Toxic Release Inventory (TRI) in Tracking Implementation and Environmental Impact of Industrial Green Chemistry Practices in the United States

Sandra Duque Gaona

Additional information is available at the end of the chapter

http://dx.doi.org/10.5772/intechopen.70716

Abstract

The Toxics Release Inventory is a rich data source with nearly 30 years of reported data from industrial facilities located in the United States. Annually, these facilities report on their chemical waste management practices, including the quantities they release to air, water, and land; treat; combust for energy recovery; or recycle. Facilities are also required to disclose any green chemistry or other pollution prevention practices, and have the option to provide additional details on these practices or on barriers they encounter in implementing such practices. The Toxics Release Inventory (TRI) provides a means by which a facility's or industry sector's implementation of green chemistry practices can be tracked, and the impact that these practices have on environmental performance. This chapter describes analytical options for tracking implementation of green chemistry practices and assessing the environmental impact of such practices. Key TRI data elements are highlighted as well as where to obtain the information.

Keywords: green chemistry, codes, source reduction, toxics, chemicals, TRI, releases, reporting

1. Introduction

Facilities that are subject to the Toxics Release Inventory (TRI) reporting requirements are required to disclose any source reduction practices implemented at their facilities during the year for which they are reporting. Facilities report the newly implemented source reduction practices by choosing one or more predefined codes (W-codes) that correspond to a specific practice within the eight established TRI source reduction categories (e.g., process modifications, substitution of raw materials).

Over the past 2 decades many facilities have implemented green chemistry practices in their operations that reduce or eliminate the use or generation of TRI-listed chemicals to prevent pollution at its source. In doing so, facilities improve their environmental performance while off-setting the continually rising costs of managing production-related toxic chemical wastes. Beginning with the 2012 TRI reporting year, in recognizing that none of the existing source reduction codes (W-codes) were specific to green chemistry, the U.S. Environmental Protection Agency (EPA) implemented six new codes to align closely with green chemistry practices (e.g., W15, introduced in-line product quality monitoring or other process analysis system and W43, substituted a feedstock or reagent chemical with a different chemical), to enable facilities to disclose adoption of these practices.

This chapter introduces the EPA's TRI program and how the TRI has evolved over the past 30 years into a pollution prevention resource. TRI data specific to source reduction will be described, followed by discussions on how these data can be used to assess industrial progress in implementing green chemistry practices and possible impacts on the reduction of TRI-listed chemical generation and releases to the environment.

2. Toxics Release Inventory (TRI) Program

The TRI program was established by the Emergency Planning and Community Right-to-Know Act (EPCRA) in 1986 [1], and TRI reporting commenced with the 1987 reporting year (first TRI reports due July 1st, 1988), and has continued to the present. The 2015 reporting year marked 29 years of available TRI data, resulting in a rich source of information on TRI-listed chemicals, which now exceeds over 650 discrete chemicals and 30 chemical categories [2].

The Pollution Prevention Act (PPA) of 1990 expanded TRI's authority to collect information beyond release quantities as specified in EPCRA Section 313 to include information specific to source reduction and preferred waste management techniques as described under Section 6607 of the PPA [3]. This change was significant giving the public a broader lens by which to evaluate and track corporate performance in their management of TRI-listed chemicals.

As illustrated in (**Figure 1**), the waste management hierarchy [4], since reporting year 1991, for a given chemical on the TRI list, facilities are *required* to report the quantities of the chemical recycled, used for energy recovery, or treated at the facility or elsewhere in addition to the original reporting requirements on releases emitted directly into the environment or transferred off-site to disposal, treatment, or storage facilities. Optional waste minimization information also transitions to a formal requirement where facilities must report any source reduction activities (e.g., process modifications, substitution of raw materials) newly implemented at the facility during the reporting year.

http://dx.doi.org/10.5772/intechopen.70716

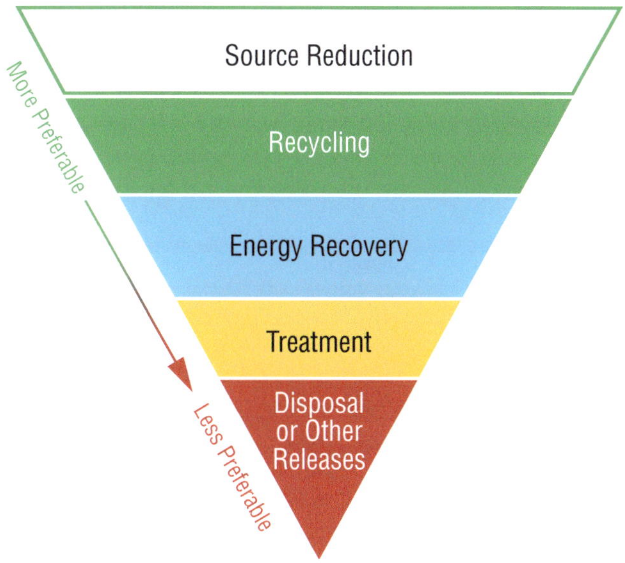

Figure 1. Waste management hierarchy.

For reporting year 2012, the TRI program, cognizant of the advancements in science and initiatives underway at facilities, expanded the codes available to facilities under the source reduction categories to better align with green chemistry practices. The addition of these six new codes is discussed in greater detail in the next section.

During this same time frame, the program made additional enhancements to the reporting form allowing facilities the option to specify barriers that were preventing them from implementing source reduction activities. Previously, facilities only had the opportunity to provide commentary without adequate data fields for tracking purposes.

Of greatest value perhaps to TRI data users are the open text data fields. Facilities (since 1991) can provide additional optional commentary to describe their source reduction activities, other environmental practices, or other activities reported to the TRI program such as reasons for increased releases. This field has the potential to be an important communication mechanism if used by industry. For this reason, the TRI encourages the submittal of optional information, for it not only augments understanding of industrial management, but provides a unique opportunity for facilities to showcase and further extend successful pollution prevention practices.

2.1. Evolution of the TRI reporting form

Facilities have had the option to report on pollution prevention activities since the start of the TRI program. For the first 4 years (1987–1990) of the program, prior to implementation of the additional TRI reporting requirements established under the PPA, facilities could voluntarily provide information on waste minimization (pollution prevention) through the selection of one of eight codes shown in **Table 1** that best described their activities. Facilities could also indicate the effect of these activities on the quantities released by providing a waste minimization index helping to distinguish between business activities and minimization efforts. However, this optional data collected in Section 8 of the TRI form was highly underreported. Note that recycling was included within this category and later separated [5].

Code	Description	Example
M1	Recycling/reuse on-site	Solvent recovery still; vapor recovery systems; reuse of materials in a process
M2	Recycling/reuse off-site	Commercial recycler; toll recycling; at an off-site company-owned facility
M3	Equipment/technology modifications	Change from solvent to mechanical stripping; modify spray systems to reduce overspray losses; install floating roofs to reduce tank emissions; install float guards to prevent tank overflow
M4	Process procedure modifications	Change production schedule to minimize equipment and feedstock change-overs; improved control of operating conditions; segregation of wastes to permit recycling
M5	Reformulation/redesign of product	Change in product specifications; modify design or composition; reduce or modify packaging
M6	Substitution of raw materials	Change or eliminate additives; substitute water-based for solvent-based coating materials, cleaners, and pigments; increase purity of raw materials
M7	Improved housekeeping, training, inventory control	After maintenance frequency; institute leak detection program; improved inventory control; institute training program on waste minimization
M8	Other waste minimization technique	Elimination of process; discontinuation of product

Table 1. Pre-PPA codes, waste minimization codes.

Recognizing the importance of this information as a possible way to address chemical wastes and operations at industrial facilities, regulators significantly expanded Section 8 of the TRI reporting form (Form R) and made mandatory the reporting of pollution prevention (P2) activities as of reporting year 1991. Source reduction activities implemented during a year would be reported through the selection of the appropriate code(s) indicating the type of actions taken to reduce chemical waste: disposed of or released, treated, used for energy recovery, or recycled. Facilities could select from the 43 codes listed in **Table 2** that correspond to eight source reduction categories [6].

The expanded Section 8 of the TRI Form R also includes other reporting requirements specified by the PPA on quantities of chemical waste managed as waste (which includes recycled, burned for energy recovery, treated, or released). This section often represents a summary of more detailed information presented in other sections, such as releases in Sections 5 and 6 or on-site treatment methods and efficiencies in Section 7. Beyond the additional report data elements, following the PPA, the reporting form was reorganized and condensed into two parts, combining previous Parts II and III into the current Part II on Chemical Information.

http://dx.doi.org/10.5772/intechopen.70716

Source reduction category	Source reduction code	Source reduction description
Good operating practices	W13	Improved maintenance scheduling, record keeping, or procedures
	W14	Changed production schedule to minimize equipment and feedstock changeovers
	W19	Other changes made in operating practices
Inventory control	W21	Instituted procedures to ensure that materials do not stay in inventory beyond shelf-life
	W22	Began to test outdated material — continue to use if still effective
	W23	Eliminated shelf-life requirements for stable materials
	W24	Instituted better labeling procedures
	W25	Instituted clearinghouse to exchange materials that would otherwise be discarded
	W29	Other changes made in inventory control
Spill and leak prevention	W31	Improved storage or stacking procedures
	W32	Improved procedures for loading, unloading, and transfer operations
	W33	Installed overflow alarms or automatic shutoff valves
	W35	Installed vapor recovery systems
	W36	Implemented inspection or monitoring program of potential spill or leak sources
	W39	Other changes made in spill and leak prevention
Raw material modifications	W41	Increased purity of raw materials
	W42	Substituted raw materials
	W49	Other raw material modifications made
Process modifications	W51	Instituted re-circulation within a process
	W52	Modified equipment, layout, or piping
	W53	Used a different process catalyst
	W54	Instituted better controls on operating bulk containers to minimize discarding of empty containers
	W55	Changed from small volume containers to bulk containers to minimize discarding of empty containers
	W58	Other process modifications made

Source reduction category	Source reduction code	Source reduction description
Cleaning and degreasing	W59	Modified stripping/cleaning equipment
	W60	Changed to mechanical stripping/cleaning devices (from solvents or other materials)
	W61	Changed to aqueous cleaners (from solvents or other materials)
	W63	Modified containment procedures for cleaning units
	W64	Improved draining procedures
	W65	Redesigned parts racks to reduce drag out
	W66	Modified or installed rinse systems
	W67	Improved rinse equipment design
	W68	Improved rinse equipment operation
	W71	Other cleaning and degreasing modifications made
Surface preparation and finishing	W72	Modified spray systems or equipment
	W73	Substituted coating materials used
	W74	Improved application techniques
	W75	Changed from spray to other system
	W78	Other surface preparation and finishing modifications made
Product modifications	W81	Changed product specifications
	W82	Modified design or composition of product
	W83	Modified packaging
	W89	Other product modifications made

Table 2. Post-PPA codes, source reduction codes.

Since 1991, the TRI Form R has been fine-tuned with smaller improvements for clarification purposes and to reduce reporting burdens. The gradual transition from 2006 to 2014 from paper form reporting to an electronic-only system, with the exception of those facilities claiming trade secret information also helped greatly with data quality and increased reporting of optional descriptive information. Moreover, significant to pollution prevention and green chemistry are the additions for the 2012 reporting year [7]. The 2012 update allows for the tracking of green chemistry activities as well as better tracking of barriers to source reduction. As explained in the introduction, six green chemistry source reduction codes were added expanding the total number of source reduction codes to 49. Noticing that facilities were providing commentary on obstacles, the TRI Program also developed eight codes that enable facilities to disclose (voluntarily) the most common barriers to source reduction implementation. These additional codes are listed in **Table 3**.

http://dx.doi.org/10.5772/intechopen.70716

Green chemistry code	Green chemistry code description	Barrier code	Barrier code description
W15	Introduced in-line product quality monitoring or other process analysis system	B1	Insufficient capital to install new source reduction equipment or implement new source reduction activities/initiatives
W43	Substituted a feedstock or reagent chemical with a different chemical	B2	Require technical information on pollution prevention techniques applicable to specific production processes
W50	Optimized reaction conditions or otherwise increased efficiency of synthesis	B3	Concern that product quality may decline as a result of source reduction
W56	Reduced or eliminated use of an organic solvent	B4	Source reduction activities were implemented but were unsuccessful
W57	Used biotechnology in manufacturing process	B5	Specific regulatory/permit burdens
W84	Developed a new chemical product to replace a previous chemical product	B6	Pollution prevention previously implemented-additional reduction does not appear technically or economically feasible
		B7	No known substitutes or alternative technologies
		B8	Other barriers

Table 3. Green chemistry and barrier codes added in 2012.

2.2. TRI data elements

For analytical purposes to track the implementation and impact of green chemistry practices, five overarching data elements are important. Background on source reduction has already been provided and to a lesser extent on optional pollution prevention (P2) text. These first two elements along with production information help understand the quantitative values (waste managed and releases) reported for TRI-listed chemicals. These elements are:

- **Optional P2 Text**, which includes narratives on P2-related activities and provide greater context for understanding source reduction activities, other environmental management practices, as well as barriers to source reduction implementation at the facility (reported in TRI Form R Section 8.11).
- **Source Reduction**, which includes newly implemented activities that reduce or eliminate the generation of pollutants (reported in TRI Form R Section 8.10). Source reduction practices include for example process modifications and substitution of raw materials.
- **Production Ratio (PR) or Activity Index (AI)**, which specifies the level of increase or decrease from the previous year, of the production process or other activity in which the toxic chemical is manufactured, processed or otherwise used (reported in TRI Form R Section 8.9). This number is usually around 1.0. For example, a production ratio or activity index of 1.5 indicates about a 50% increase in production from the prior year associated with, for example, the use of the chemical, while a value of 0.3 indicates about a 70% decrease in production associated with the chemical.

- **Waste Managed**, which includes all quantities of waste that are recycled, used for energy recovery, treated, or released whether on-site or transferred off-site (reported in TRI Form R Sections 8.1 through 8.7). Waste managed tracks production-related waste only and does not include quantities associated with accidental or remedial one-time events.
- **Releases**, which includes all quantities disposed of or otherwise released to the environment through all release mechanisms to all media, whether on-site or transferred off-site to a publically owned treatment works (POTWs) or other facility for disposal, treatment, or storage (reported in TRI Form R Sections 5 and 6). Release quantities track both production and non-production related releases. Releases to air include stack and fugitive emissions. Releases to land include, for example, disposal in landfills and injection into underground wells. Releases to water include discharges into rivers, streams, or other bodies of water.

2.3. TRI data reporting and access

Before delving into analytical methodologies, it's important to understand the segment of industrial activity that TRI covers. TRI represents a slice of industrial activity. The inventory collects information from larger industrial facilities that meet the TRI reporting criteria for the employee threshold, the chemical manufacture, processing or otherwise use threshold, and operate within an industry covered sector. Specifically, facilities are subject to reporting if they (1) have ten or more full-time employees, (2) are in a TRI-covered industry NAICS code such as the manufacturing sector and other sectors (e.g., electric utilities, metal mining, and hazardous waste management) or are federal facilities, and (3) manufacture or process more than 25,000 lb., or otherwise uses more than 10,000 lb. of a TRI-listed chemical within a calendar year. Thresholds for persistent bioaccumulative toxic (PBT) chemicals are lower – as low as 0.1 g for dioxin – due to their potentially greater threat to human and environmental health.

Facilities subject to the TRI reporting requirements report annually by July 1st of each year to EPA's TRI Program, and state and tribal governments [8]. Each year, EPA's TRI Program receives approximately 80,000 form reports from approximately 20,000 facilities [9]. Form reports are chemical and chemical category specific and facilities that exceed the thresholds discussed above for a specific calendar year are required to report on the data elements outlined above as well as others.

EPA makes this information available and readily accessible to the public through various data tools, maintained by EPA's TRI Program. Various access options are discussed later in the chapter.

2.4. Analytical considerations and methodologies

In order to conduct sound analysis of green chemistry activities reported to the TRI Program, certain considerations are key for tailoring the research. Three considerations are outlined below using the data that can be derived from the TRI dataset.

Tracking a set of facilities: Analysis of the reported quantities for waste managed and released in the year the source reduction activity was reported may not lead to any significant insight as implementation of an action may not result in immediate effects. Therefore, instead of gathering data for the specific years associated with green chemistry codes, set analyses are recommended. For example, to fully understand the potential impact of green chemistry practices it is important to track the set of facilities that reported green chemistry for specific chemicals over a broad time frame. Gathering pre-source reduction quantities as well as post-source reduction quantities would give some insight as to the impact of the change.

Production levels: Consideration should also be given to production information and whether the facility is operating within normal ranges and not below or above for the time span being considered. The reported production ratio or activity index help understand the quantitative values reported and assess whether changes (increases or decreases) are due to shifts in production levels or attributable to other factors such as the implementation of new source reduction activities. Increasing or stable production coupled with decreasing releases is a positive indicator of effective pollution prevention practices.

Focus on subgroups: To more profoundly understand the magnitude of the impact, segmenting the data by industry (e.g., specific industry sector or subsector) can inform on overall activities undertaken by similar businesses. Facilities reporting to the TRI can specify up to six North American Industrial Classification System (NAICS) codes with one as the primary NAICS code, corresponding to their primary business activity. More in-depth analysis using industry-chemical combinations may also be advantageous to more accurately assess green chemistry impacts of certain chemicals or types of chemicals. Geographic analysis as an additional layer to the industry studies or as a separate subgroup option may provide some insight on local policies or clustering of mutually-beneficial resources.

The TRI dataset, while very comprehensive as a multi-media inventory of releases and other waste management information, should not be studied in isolation. Consideration of TRI in conjunction with other data sources will allow for more holistic assessment of green chemistry impacts in light of other confounding factors. For example, external factors such as outsourcing (transferring manufacturing and production operations to facilities in other countries) and the state of the economy should also be evaluated. A study published in 2015 considering this same topic of assessing the implementation and effectiveness of green chemistry in industrial manufacture of chemicals, but focused on TRI and pharmaceutical manufacturers, describes how these external factors can be considered [10]. Another valuable resource that discusses more general details on limitations of the TRI data is EPA's document on *Factors to Consider when using TRI Data* [11].

3. Tracking implementation of source reduction and green chemistry in the US

3.1. Source reduction

According to the Pollution Prevention Act of 1990, source reduction is any practice that:

> *"reduces the amount of any hazardous substance, pollutant, or contaminant entering any waste stream or otherwise released into the environment (including fugitive emissions) prior to recycling, treatment, or disposal; and reduces the hazards to public health and the environment associated with the release of such substances, pollutants, or contaminants."*

Pollution can be reduced at its source by a wide variety of techniques, prior to end-of-pipe pollution controls or recycling, such as by changing the product, materials, or processes that generate pollution in the first place. Because of the potential advantages of these preventative approaches, the U.S. EPA took steps to encourage industrial facilities to engage in source reduction. On their part, industrial facilities have engaged in substantial pollution prevention efforts, by carrying out 447,000 unique source reduction activities between 1991 and 2015 (as reported to the EPA's TRI Program).[1] **Figure 2** shows that many facilities (about 107,000) conducted these source reduction projects over the past 25 years [12].

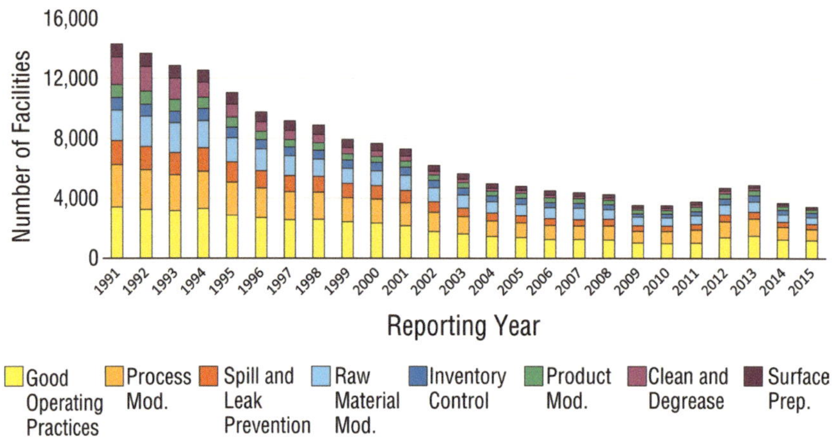

Figure 2. Facilities with source reduction projects.

Based on the eight source reduction categories tracked, the trend graph above shows that the most reported source reduction category is good operating practices. Source reduction data reported for 2015 (**Figure 3**) show that good operating practices represents 40% followed by the process modifications category at 21%. The two least reported categories are surface preparation and finishing as well as cleaning and degreasing.

3.2. Green chemistry

According to the U.S. EPA, "Green chemistry is the design of chemical products and processes that reduce or eliminate the use or generation of hazardous substances. Green chemistry applies across the life cycle of a chemical product, including its design, manufacture, use, and ultimate disposal" [13].

There are many benefits to implementing green chemistry that are inextricably linked to its preventative premise. These include improved economy and business, environment, and human health conditions.

[1]The results have been updated from previously published results (Ranson et al. [16]) to include the 2013 to 2015 TRI data.

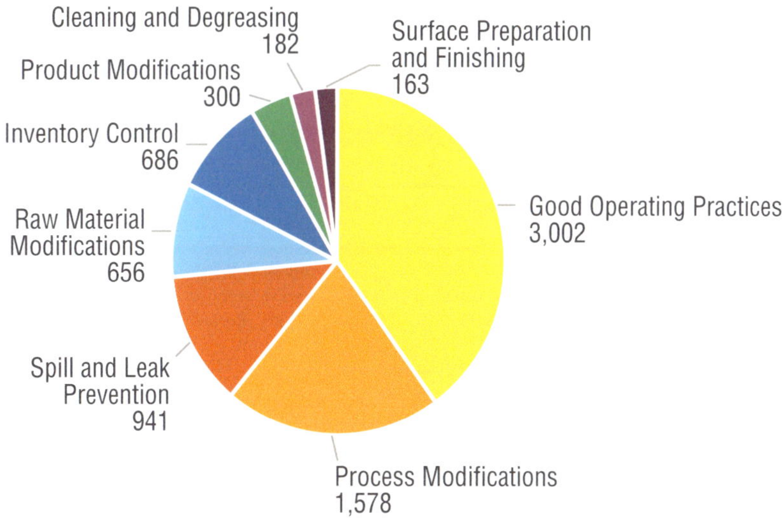

Figure 3. Number of source reduction activities.

Economy and business benefit from reduced waste generation, eliminating costly remediation in the event of accidental releases, hazardous waste disposal, and end-of-the-pipe treatments. Implementing green chemistry saves money by offsetting the costs associated with managing toxic or hazardous chemical waste. In terms of the chemicals, it reduces the need and demand for the manufacture of TRI-listed chemicals while incentivizing the creation of less toxic or non-toxic chemicals, improving competitiveness of chemical manufacturers and their customers. Use of green chemistry and associated safer-product labeling (e.g. Safer Choice labeling) [14] may also lead to increased consumer sales (by earnings).

The environment benefits from reduced emissions of TRI-listed chemicals or other hazardous substances, signifying less chemical disruptions to ecosystems. Through green chemistry, the environment would benefit from reductions in emissions of toxics to air, water, and land such as reduced use of landfills, especially hazardous waste landfills. Plants and animals also suffer less harm from reductions in hazardous chemicals entering the environment.

Human health also benefits from cleaner environmental conditions. Cleaner air resulting from reductions in hazardous chemicals released to air leads to reduced respiratory disease and other illnesses. Similarly, cleaner water resulting from reductions in hazardous chemicals released to water lead to cleaner drinking and recreational water. Application of green chemistry results in safer consumer products that enter the market and are available for purchase, thereby increasing the safety of consumers and society in general. These products may be new, replacements for less safe products (e.g., certain pesticides, cleaning products), or designed to be manufactured efficiently and with less accompanying waste (e.g., drugs). This preventive practice also benefits the workers in the chemical industry resulting in increased safety through less use of toxic materials, reduced potential for exposure and accidents (e.g., fires or explosions), and reduced need for personal protective equipment.

Given these benefits, it is not surprising to see industry advances in green chemistry. In 2012, the TRI program added six green chemistry source reduction codes to better track these ongoing activities and their possible improvements. These codes are captured within 4 of the 8 categories and are listed in (**Table 4**) along with guidance provided to reporters for increased data quality [15].

Source reduction categories	Green chemistry codes	Guidance in TRI reporting forms
Good operating practices	**W15:** Introduced in-line product quality monitoring or other process analysis system	Select code W15 if the introduction of such a system led to a reduction in the amount of the EPCRA Section 313 chemical generated as waste.
Raw material modifications	**W43:** Substituted a feedstock or reagent chemical with a different chemical	Select code W43 if the EPCRA Section 313 chemical was a feedstock or reagent chemical and you replaced it (in whole or in part) with a different chemical. • For raw material substitutions not at the level of the individual chemical (e.g., the substitution of natural gas for coal), select instead W42 *Substituted raw materials.* • If use of a feedstock or reagent chemical was reduced or eliminated because of a change in the final product, select instead one of the codes listed under *Product Modifications.*
Process modifications	**W50:** Optimized reaction conditions or otherwise increased efficiency of synthesis **W56:** Reduced or eliminated use of an organic solvent **W57:** Used biotechnology in manufacturing process	Select code W50 if the amount of the EPCRA Section 313 chemical generated as waste was reduced by increasing the overall efficiency of the synthesis. • If efficiency of syntheses was improved by using of a different catalyst, select instead W53 Used a different process catalyst. Select code W56 if the EPCRA Section 313 chemical was used as a solvent in the process and the process was modified such that the EPCRA Section 313 chemical was either replaced or no longer used in as large a quantity. Select code W57 if the use of biotechnology in the process reduced or eliminated the use of the TRI chemical.
Production modifications	**W84:** Developed a new chemical product to replace a previous chemical product	Select code W84 if the EPCRA Section 313 chemical had been produced at the facility but was replaced it (in whole or in part) with the production of a different chemical or chemicals.

Table 4. Green chemistry codes and reporting guidance.

3.2.1. Tracking green chemistry by year and code

From 2012 to 2015, TRI reporting rates by year and code show that of the 37,117 total source reduction activities reported, 1756 (5.1%) were reported as green chemistry (i.e., reported on a Form R using one of the six green chemistry codes). The vast majority were reported as W15 or W50 as shown in **Figure 4**. These relatively high reporting rates indicate that facilities are seizing opportunities for increased monitoring and efficiencies. Whereas a minimum number of facilities reported W57, demonstrating limited implementation of biotechnology in manufacturing processes.

http://dx.doi.org/10.5772/intechopen.70716

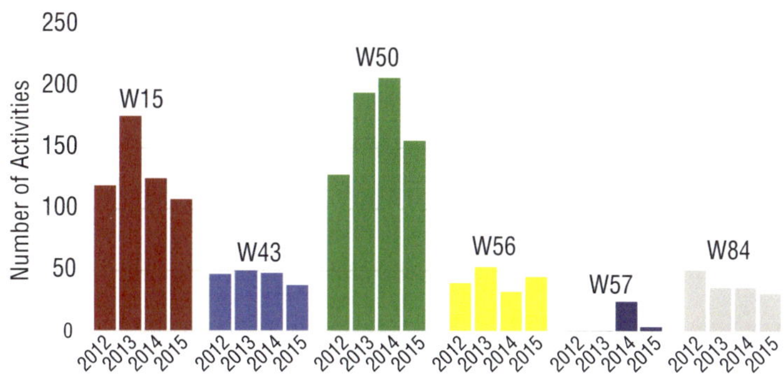

Figure 4. Green chemistry by year and code.

3.2.2. *Tracking green chemistry by industry sector*

On an industry sector level, implementation of green chemistry and total source reduction activities reported from 2012 to 2015 is visible for the top six sectors shown in **Figure 5**. The chemical manufacturing industry makes up the greatest percentage of all green chemistry reporting and constitutes a greater percentage of green chemistry reporting than of total source reduction reporting for the sector (35% vs. 29%). Both metrics are consistent with Fabricated Metal Product Manufacturing in second place, respectively at 13 and 12%. Differences in industry reporting are notable at the third level and beyond with the following observations:

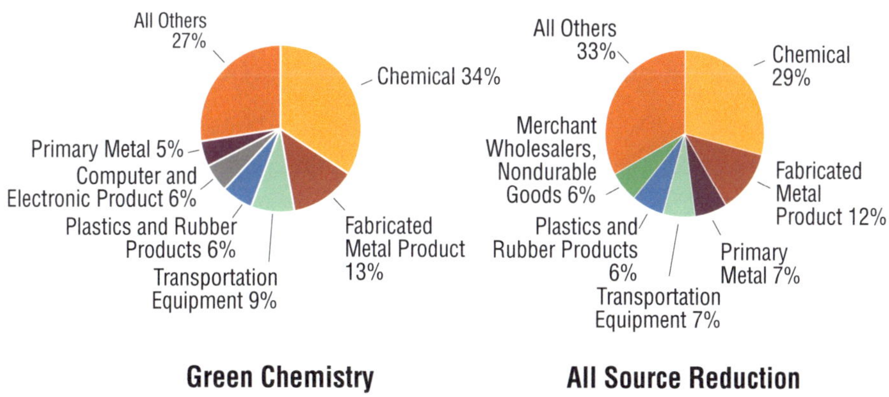

Figure 5. Green chemistry versus total source reduction by sector, 2012–2015.

- The Primary Metal Manufacturing dropping from 7% for all source reduction activities to nearly 5% for only green chemistry. A possible reason for this may be that the nature of the business may not be as amenable to green chemistry as it is in the chemical manufacturing industries.
- The Transportation Equipment Manufacturing sector covers the Automotive Manufacturing sector (NAICS 3361-3363) and as expected given recent advances, the majority (70%) of green chemistry reporting is from the auto sector. Overall, the transportation sector represents a larger share of green chemistry reporting compared to total source reduction reporting (9% vs. 7%).

- The Merchants and Wholesalers sector, while actively implementing source reduction activities and within the top six, is almost nil for ranking based on green chemistry with 0.2% representing three activities during the 4-year time period.
- The Computer and Electronic Product Manufacturing sector, while not delineated in the source reduction pie chart, falls in seventh position representing 5% of the "all others" category. This indicates that the computer manufacturing sector implemented a consistent share of green chemistry activities to source reduction activities.

How does reporting of green chemistry implementation compare to all TRI reporting? Tracking the implementation of green chemistry in the context of all TRI reports is important because it provides a lens as to sectors more amenable to green chemistry practices and where collaborative efforts may be more readily established. High TRI reporting rates from sectors that do not report green chemistry practices are likely indicators that such sectors face source reduction obstacles. Barriers are discussed in more detail later in the chapter. The pie charts in **Figure 6** show that three of the six sectors fall within the top ranking for both green chemistry and overall TRI reporting. Other sectors such as Petroleum and Coal Products Manufacturing and Utilities, while high in number of TRI forms submitted to TRI, do not report many source reduction activities or specific green chemistry practices for TRI-listed chemicals.

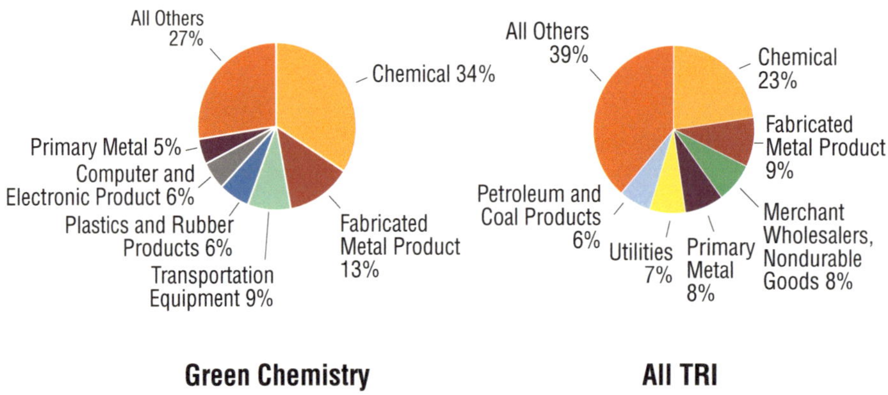

Figure 6. Green chemistry versus TRI reporting by sector, 2012–2015.

More in-depth analysis by NAICS code is recommended to help delineate more precisely green chemistry implementation by facilities within specific subsectors of a given industrial sector and their environmental impact. For example, the case study involving TRI and Pharmaceutical Manufacturers to assess the implementation and effectiveness of green chemistry practices focused on facilities classified in NAICS codes 325411 (Medicinal and Botanical Manufacturing) and 325412 (Pharmaceutical Preparation Manufacturing) [10]. This sector represents 16.5% of the chemical manufacturing sector or about 6% of all industry sectors that reported green chemistry practices to TRI from 2012 to 2015.

Case Study Focus: The study examined TRI data submitted for reporting years 2002 through 2011 and, hence did not include consideration of the green chemistry codes since the codes were implemented for reporting year 2012. Nonetheless, the analyses show that over the 2002–2011 timeframe the quantities of TRI chemicals reported annually by pharmaceutical manufacturing facilities to EPA's TRI Program as released to the environment or otherwise managed as waste declined steadily and by more than 60%. The downward trend was largely driven by reductions in the quantities reported for organic solvents. Five solvents (methanol, dichloromethane, toluene, dimethylformamide and acetonitrile) accounted for three-quarters of the declining trend in production-related waste managed, which includes environmental releases. Overall, the reductions in reported quantities are sector-wide, and it appears that factors such as outsourcing, production levels, regulations, shifts to other waste management practices, or larger pharmaceutical firms did not precipitate the decline. The authors concluded from their analyses and the extensive evidence in the literature of green chemistry advances within the pharma sector that implementation of green chemistry practices is a major contributing factor to the large reductions [10].

3.2.3. *Tracking green chemistry by chemical*

Green chemistry implementation can also be tracked on a chemical level. Industrial facilities reported green chemistry activities to reduce the generation of waste of the following chemicals. **Figure 7** shows the top 8 chemicals based on total green chemistry reporting from 2012 to 2015 and delineates the individual green chemistry codes selected. The majority of green chemistry codes were reported for methanol, toluene, copper, and ammonia, representing 21%. With the chemical manufacturing industry ranking first and the published solvent reduction advancements, TRI data confirm industries' efforts to implement projects to reduce methanol and toluene, the top two most reported chemicals [10]. The top W-codes selected were W50, optimized reaction conditions, followed by W15, in-line product quality monitoring.

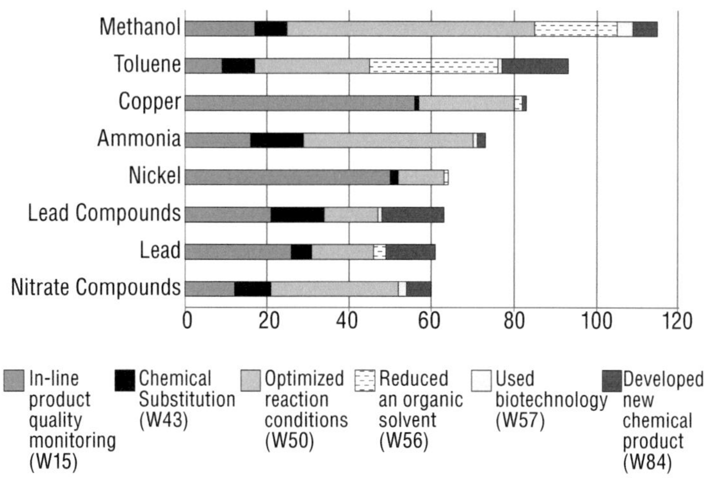

Figure 7. Green chemistry by chemical.

3.3. Assessing impact of industrial green chemistry practices

In practice, implementing source reduction activities aims to improve environmental performance, and as TRI-listed chemicals are eliminated or reduced in processes, facilities consequently reduce associated costs with managing production-related waste of those chemicals. However, what do the data indicate? Do the data confirm that implementation of green chemistry techniques results in reduced waste management and release quantities?

Based on a previous statistical analysis using the "differences-in-differences" methodology, all implemented source reduction is not equal, meaning all activities do not equally decrease the quantities of chemical waste managed. The study, which considered a wide range of TRI data from 1987 to 2012, shows that there is considerable variation in how the implementation of different source reduction activities affects releases. For example, good operating practices, which is the category corresponding to green chemistry code W15, has only a small effect (roughly −4%). In contrast, source reduction activities focused on raw material modifications, which contains green chemistry code W53, shows a large decrease in releases of −20%. Similarly, product modifications, including W84 shows a −13% decrease. The other green chemistry codes fall under the process modifications category, which has shown moderate decreases of −5% [16]. One can infer from this study that to quantify the effectiveness of source reduction, different green chemistry practices would result in different environmental impacts.

This study also shows that impacts may be experienced up to 5 years following the implementation of a source reduction project. Conducting a similar type of analysis focused on green chemistry practices, especially now that codes are available to clearly track any associated projects would serve as a good case study to verify the overall results. However, additional data is needed to apply this methodology and conduct a robust statistical analysis to observe the long-term impact of green chemistry practices. Within 3–5 years from the time of this writing, sufficient data will be available to evaluate the effectiveness of those activities implemented from 2012 to 2015. As mentioned previously, tracking the same set of facilities over time will ensure visibility of any impacts associated with green chemistry approaches.

3.3.1. Impact of green chemistry on waste managed quantities

Analysis of the green chemistry practices implemented during 2012 and the impact these practices had on the quantities of TRI chemical waste managed is presented below. To account for at least one factor that could influence changes in the quantities of chemical waste managed, the analysis normalizes based on reported production values. Considering only those facilities that reported green chemistry codes for 2012 and reported production ratios within the normal range (greater than 0.2 and less than 3) and consistently for all years in the time span, the normalized production-related waste managed trend in **Figure 8** shows 7 years of data with 3 years prior to 2012 and 3 years after 2012. The decrease in waste managed during 2012 indicates that green chemistry actions implemented during that year could have contributed to the observed reduction. Investigation into the release quantities for 2013 and 2014 indicates that two facilities are primarily responsible for increases in releases and treatment.

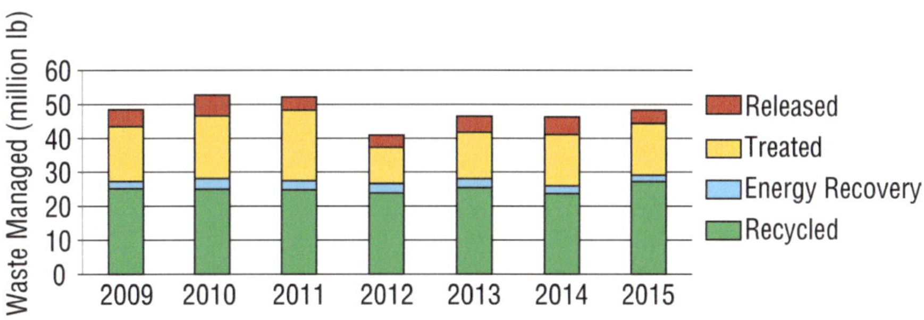

Figure 8. Production normalized waste managed, 2009–2015.

BASF CORP-SAVANNAH OPERATIONS, TRIFID 31404KTLST1800E, NAICS 327992: Ground or Treated Mineral and Earth Manufacturing. Facility reported green chemistry code, W50, for nitrate compounds for 2012. The 3 years following show highest releases for 2014, with 2,860,000 pounds (55% of total releases) discharged to water.

ARKEMA INC CLEAR LAKE, TRIFID 77507DWCHM952BB, NAICS 325110: Petrochemical Manufacturing. Facility reported green chemistry code, W50, for two chemicals: butyl acrylate and n-butyl alcohol for 2012. The years following show the highest treatment quantities of butyl acrylate for 2013 with 3,848,260 pounds, 28% of the total treated waste. For n-butyl alcohol, 1,732,045 pounds were treated during 2014, representing 11% of total treated waste during the year.

The formula used to calculate the normalized trend is as follows. It is applied to all year-facility-chemical combinations to obtain a normalized production value for each. Year 2009, as the first year in the series, is set as the base year equal to 1.

P = absolute production

PR = production ratio (provided by facility. relative to previous year.)

PI = production index relative to 2009

W = absolute waste quantity

PNW = production normalized waste quantity

General formulas:

$$PIyear\ x = PIyear\ x - 1 * PRyear\ x \tag{1}$$

$$PNWyear\ X = \frac{Wyear\ X}{PIyear\ X} = \frac{Wyear\ X}{PIyear\ X - 1 * PRyear\ X - 1} \tag{2}$$

Example 1:

$PI_{2009} = 1$

$PNW_{2009} = W_{2009}$

Example 2:

$$PI_{2010} = PI_{2009}{*}PR_{2010} = PR_{2010}$$

$$PNW2010 = W2010/PI2010 = W2010/PR2010$$

Example 3:

$$PI_{2011} = PI_{2010}{*}PR_{2011} = PR_{2009}{*}PR_{2010}{*}PR_{2011}$$

$$PNW2011 = W2011/PI2011 = W2011/(PR\,2010^{*}\,PR2011)$$

A more direct analysis of the data without consideration of production indicates implementation of green chemistry practices as favorable to lowering waste management quantities. Comparing the 2012 subset of facilities that reported green chemistry codes to all other facilities that reported to the TRI Program for the same year, shows that facilities reporting green chemistry have a larger decrease in their waste managed compared to all facilities. Out of 249 facilities that reported implementation of a green chemistry practice during 2012, 59.2% of those facilities decreased their waste from 2011 to 2015. While 47.6% of facilities that did not report implementation of a green chemistry practice during 2012 decreased their waste from 2011 to 2015.

Assessing impact is both a beneficial exercise and a difficult one because facilities do not directly report the extent to which green chemistry impacts production-related waste managed. However, the optional text that facilities may include in their reports does provide additional insight as to the specific practices implemented and their success. As an example, Cathay Industries USA Inc., in Valparaiso, Indiana in the Chemicals Manufacturing sector, Synthetic Dye and Pigment Manufacturing (NAICS 325130), reported green chemistry code, W50, "Optimized reaction conditions or otherwise increased efficiency of synthesis" for both 2012 and 2013 for ammonia. Normalized production waste management trends of ammonia show decreases in those years, and continued low levels in 2014–2015. Additionally, Cathay Industries noted "Improved measurement and control of reactant / reaction" in the source reduction optional text field for the Form R filed for reporting year 2013 [17]. This additional context could be useful for encouraging similar best practices at other facilities.

More focused analysis by industry sector or green chemistry code would provide more insightful findings as well as more accurate estimates of impact. Analysis of waste managed quantities help to track the overall performance of the facility and more granular analysis of each of the waste management methods, particularly the releases portion, which would inform on progress toward reducing the emission of toxic chemicals to environmental media.

4. Accessing TRI green chemistry data

Over the time span of the TRI program various tools for accessing and analyzing TRI data have been developed comprising the TRI tool suite available today. Depending on data user objectives, some tools are better suited for certain purposes than others. Three resources are described below and summarized in **Table 5**.

http://dx.doi.org/10.5772/intechopen.70716

TRI data resources	Description
TRI Pollution Prevention Search Tool	Easiest method to explore and access P2 related information by facility and conduct comparisons on an industry scale
TRI National Analysis Supporting Data File "Additional P2 Data" Download	Pre-formatted downloadable P2 data file for a specific reporting year
TRI Customized Search Tool	Most robust tool for ad-hoc querying of all TRI reported data fields

Table 5. TRI data resources to access green chemistry data.

In the realm of pollution prevention data, the best way to explore all available P2 information is through **TRI's Pollution Prevention Search Tool** [18]. A user can easily query for all information reported for a specific year or can further limit to a specific chemical or industry. The results table shows the source reductions codes reported along with any optional text. The P2 text filter box can be adjusted to display all comments. This data can then be downloaded and more easily filtered to show only those facilities that reported green chemistry. The P2 tool is also a great way to explore the data on a facility level or to compare to other industries. For general instructions on how to conduct an industry analysis using the P2 tool, see the How-to Guidance [19].

For downloading a comprehensive set of P2 data per reporting year, the **TRI National Analysis supporting data files** are a good resource. Refer to the file "Additional P2 Data" [20]. A quick link is available from TRI's P2 webpage [4] or can be obtained directly from the National Analysis Download Report tab. The Excel workbook packages the P2 data used for EPA's interpretation of the data for the given year's National Analysis report. It is a well-organized workbook with P2 data presented over several tabs including a dedicated tab on 8.10 entries (source reduction codes reported). These codes can be filtered to those specific to green chemistry.

The most robust option to download all possible data fields associated with all facilities that reported green chemistry is **TRI's Customized Search Tool** [21]. This tool provides access to all publicly available TRI reported fields and can be tailored to your data needs. The most comprehensive table is the "flat" view (v_tri_form_r_ez) and can be selected along with other tables.

5. Conclusion

This chapter describes the utility of the TRI as a useful tool for measuring the impact of green chemistry practices on reducing releases and other waste management quantities of chemicals reportable to the TRI Program, and assessing progress toward sustainability goals. As discussed, the TRI is uniquely well-suited for assessing the progress made by specific industry sectors or specific facilities therein in implementing green chemistry practices. Green chemistry codes as a new data field will become richer with time allowing for more comprehensive analysis of impact. Three to four more years of data will be especially valuable for trend analysis and longer-term assessment of effectiveness. The TRI will continue to be an excellent source for gauging progress toward sustainability as well as for promoting possible alternatives to the manufacture, processing, or use of TRI-listed chemicals.

Disclaimer

This chapter was prepared by Sandra D. Gaona of the United States Environmental Protection Agency. The contents of this chapter do not necessarily reflect the views, rules or policies of the United States Environmental Protection Agency, nor does mention of any chemical substance necessarily constitute Agency endorsement or recommendation for use. In addition, mention of any companies does not necessarily constitute Agency endorsement.

Author details

Sandra Duque Gaona

Address all correspondence to: gaona.sandra@epa.gov

United States Environmental Protection Agency, Washington, DC, USA

References

[1] United States Code Title 42. §11001 et seq. Emergency Planning and Community Right-to-Know [Internet]. 1986. Available from: https://www.gpo.gov/fdsys/pkg/USCODE-2011-title42/html/USCODE-2011-title42-chap116.htm, 2011 Edition

[2] U.S. Environmental Protection Agency. Toxics Release Inventory Program. TRI-Listed Chemicals [Internet]. https://www.epa.gov/toxics-release-inventory-tri-program/tri-listed-chemicals

[3] United States Code Title 42. §13101 et seq. Pollution Prevention [Internet]. 1990. Available from: https://www.gpo.gov/fdsys/pkg/USCODE-1999-title42/pdf/USCODE-1999-title42-chap133.pdf

[4] U.S. Environmental Protection Agency. Toxics Release Inventory Program. Pollution Prevention (P2) and TRI [Internet]. https://www.epa.gov/toxics-release-inventory-tri-program/pollution-prevention-p2-and-tri

[5] U.S. Environmental Protection Agency. Toxic Chemical Release Inventory Reporting Package for 1990. [January 1991] EPA 560/4-91-001

[6] U.S. Environmental Protection Agency. Toxic Chemical Release Inventory Reporting Form R and Instructions. Revised 1991 Version. [May 1992] EPA 700-K-92-002

[7] U.S. Environmental Protection Agency. Toxics Release Inventory Reporting Forms and Instructions. Revised 2012 Version. February 2013. EPA 260-R-13-001. Available from: https://www.epa.gov/sites/production/files/documents/ry2012rfi.pdf

[8] U.S. Environmental Protection Agency. Toxics Release Inventory (TRI) Program. Basics of TRI Reporting [Internet]. https://www.epa.gov/toxics-release-inventory-tri-program/basics-tri-reporting

[9] U.S. Environmental Protection Agency. Toxics Release Inventory (TRI) National Analysis [Internet]. January 2017. https://www.epa.gov/trinationalanalysis. 2015 TRI National Analysis

[10] DeVito SC, Keenan C, Lazarus D. Can pollutant release and transfer registers (PRTRs) be used to assess implementation and effectiveness of green chemistry practices? A case study involving the Toxics Release Inventory (TRI) and pharmaceutical manufacturers. Green Chemistry. 2015;**17**:2679-2692

[11] U.S. Environmental Protection Agency. Factors to Consider When Using TRI Data [Internet]. 2015. Available from: https://www.epa.gov/sites/production/files/2015-06/documents/factors_to_consider_6.15.15_final.pdf

[12] U.S. Environmental Protection Agency. Toxics Release Inventory- 2015 National Analysis Dataset. Retrieved February 2017 from: https://www.epa.gov/toxics-release-inventory-tri-program/download-trinet

[13] U.S. Environmental Protection Agency. Basics of Green Chemistry [Internet]. https://www.epa.gov/greenchemistry/basics-green-chemistry#definition

[14] U.S. Environmental Protection Agency. Safer Choice. Learn about the Safer Choice Label [Internet]. Available from: https://www.epa.gov/saferchoice/learn-about-safer-choice-label

[15] U.S. Environmental Protection Agency. Toxics Release Inventory Reporting Forms and Instructions. Revised 2015 Version. December 2015. EPA 260-R-15-001. Available from: https://www.epa.gov/sites/production/files/2016-01/documents/ry_2015_tri_reporting_forms_and_instructions.pdf

[16] Ranson M, Cox B, Keenan C, Teitelbaum D. The impact of pollution prevention on toxic environmental releases from U.S. manufacturing facilities. Environmental Science & Technology. 2015;**49**(21):12951-12957

[17] U.S. Environmental Protection Agency. Envirofacts. TRI Pollution Prevention (P2) Search Tool. Cathay Industries Inc.; Ammonia profile. Available from: https://oaspub.epa.gov/enviro/P2_EF_Query.p2_report?FacilityId=46383PFZRP4901E&ChemicalId=007664417&ReportingYear=2014&DocCtrlNum=&Opt=0

[18] U.S. Environmental Protection Agency. Envirofacts. TRI Pollution Prevention (P2) Search Tool. Available from: https://www3.epa.gov/enviro/facts/tri/p2.html

[19] U.S. Environmental Protection Agency. TRI Pollution Prevention Resources [Internet]. Available from: https://www.epa.gov/toxics-release-inventory-tri-program/tri-pollution-prevention-p2-resources

[20] U.S. Environmental Protection Agency. TRI National Analysis. Supporting Data Files for the 2015 National Analysis [Internet]. Available from: https://www.epa.gov/trinationalanalysis/supporting-data-files-2015-tri-national-analysis

[21] U.S. Environmental Protection Agency. Envirofacts. TRI Customized Search Tool. Available from: https://www.epa.gov/enviro/tri-customized-searchTRIcustomized

Section 6

Green Approach in Click Chemistry

Chapter 9

Green Approach in Click Chemistry

Sachin P. Shirame and Raghunath B. Bhosale

Additional information is available at the end of the chapter

http://dx.doi.org/10.5772/intechopen.72928

Abstract

The aim of the topic on click chemistry is used to synthesize various derivatives of 1, 2, 3-triazol-1-yl piperazine, 1, 2, 3-triazol-1-yl quinoxaline, one pot 1,2,3-triazole and bistriazole. These various synthesized compounds were biologically active such as antimicrobial, anti-oxidant, anticancer, antiviral, anti HIV and antitubercular activates. The heterocyclic compounds which are pharmacological active were synthesized by the Cu (I)-catalyzed Huisgen 1, 3-dipolar cycloaddition is a major example based on the click chemistry philosophy. The click chemistry in a broad sense is about using easier reactions to make compounds for certain functions of drugs. The click chemistry used as a green synthesis, because it allows the basic principles of green chemistry given by Anastas and Warner.

Keywords: click reaction, sodium ascorbate, cupper sulphate, water, room temperature

1. Introduction

In 2001, a Nobel Prize winner K. Barry Sharpless published a landmark review describing a new strategy for organic chemistry [1]. Click reaction advantages of that are high yielding, wide in scope, create only inoffensive byproducts that can be removed without chromatography, are stereo specific, simple to perform, and can be conducted in easily removable or benign solvents.

The environmentally amiable route to carbon-hetero atom bond formation, described as a click chemistry, has become known as a fast, modular, wide in scope, efficient, reliable, simple to perform to the synthesis of novel compounds with desired functionalities [2]. The name "Click Chemistry" was coined to describe this guiding principle – a principle born to meet the demands of modern day chemistry. Among the listed click reactions, Huisgen 1, 3-dipolar cycloaddition between an azide and alkyne have been widely explored due to, among others, its efficiency, versatility and inertness toward other functional groups.

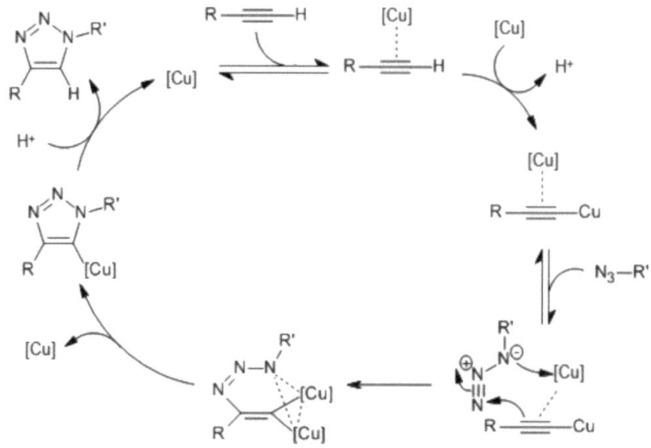

Mechanism of the click Chemistry

Rulf Huisgen reported that copper (I) salts were able to accelerate the rate of reaction. More importantly, at room temperature or at moderate temperature, the copper catalyst directs the formation of only one of them regioisomers, the 1, 4-disubstituted as shown in [3, 4].

1.1. Huisgen 1, 3-dipolar cycloaddition reaction

Rolf Huisgen, is a German chemist his major achievements was the development of the 1,3-dipolar cycloaddition reaction, also known as the Huisgen cycloaddition or Huisgen reaction. The Huisgen 1, 3-dipolar cycloaddition reaction of organic azides and alkynes, has gained considerable attention in recent years due to the introduction in 2001 of Cu (I)-catalysis by Sharpless, a major improvement in both reaction rate and chemoselectivity of the reaction, as realized by the Meldal and the Sharpless laboratories. The great success of the Cu (I)-catalyzed reaction is a quantitative, very robust, insensitive, general and orthogonal ligation reaction and use for even bio-molecular ligation [5].

1.2. Importance of Huisgen 1, 3-dipolar cycloaddition

Thermodynamic and kinetically favorable (50 and 26 kcal/mol, respectively), Regiospecific, Chemoselective, 10^7 rate enhancement over non-catalyzed reaction and triazole stable to oxidation and acid hydrolysis

One pot reactions are reactions where three or more substrates combine in one step to give a product that contains essential parts of all of them [6]. The idea of using a one pot reactions followed by a Huisgen [3+2] copper catalyzed reaction was first presented by Barbas and coworkers.

1.3. Synthesis of 1, 4-disubstituted 1, 2, 3-triazoles with copper catalyst

The copper (I)-catalyzed union of terminal alkynes and organic azides to give 1, 4-disubstituted 1, 2, 3-triazoles (as shown in **Figure 1**) exhibits remarkably broad scope and exquisite selectivity. Particularly, 1, 4-disubstituted 1, 2, 3-triazole fragment exhibit is useful for potent biological properties.

http://dx.doi.org/10.5772/intechopen.72928

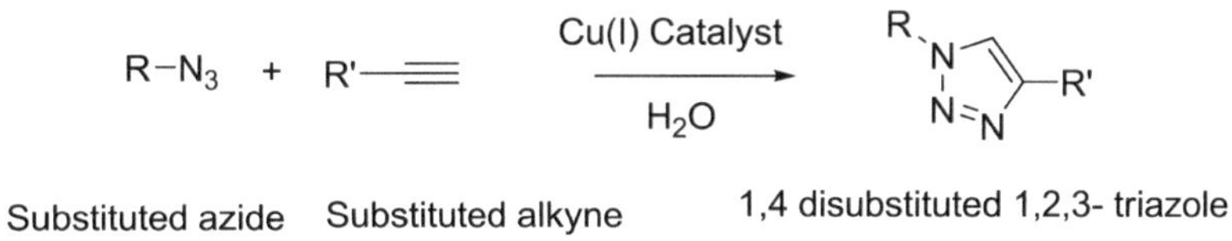

Figure 1. Copper catalyzed azide-alkyne cycloaddition.

1.4. Significance of Cu (I)-catalyzed for azide-alkyne cycloaddition reaction

Rulf Huisgen 1, 3-dipolar cycloaddition reaction is the copper (I)-catalyzed in which organic azides and terminal alkynes are combined to form 1, 4-regioisomers of 1, 2, 3-triazoles as sole products. This reaction is better termed the Copper (I)-catalyzed Azide-Alkyne Cycloaddition (CuAAC). The Cu (I) species may either be introduced as preformed complexes, or are otherwise generated in the one pot reaction itself by one of the following ways: A copper compound is present in the (+2) oxidation state is added to the reaction in presence of a reducing agent of sodium ascorbate which reduces the Cu from the (+2) oxidation state transfer to the (+1) oxidation state.

1.5. Synthesis of 1, 5-disubstututed 1, 2, 3-triazoles without copper catalyst

There are reported that the formation of regioselective 1, 5-disubstituted triazoles (as shown in **Figure 2**) being mediated by Torne [7, 8], or stereoelectronic effects, but only under harsh conditions. The chemoselectivity of the 1, 3-dipolar cycloaddition enables a convergent synthetic route to the requisite triazole.

The cycloaddition was affected by the catalyst RuCl (PPh_3), which had been reported to be effective for the cycloaddition of secondary azides. The disadvantages of 1, 5-disubstituted 1, 2, 3-triazole are not stable, so we have selected 1, 4-disubstituted 1, 2, 3-triazole.

1.6. Click chemistry acts as a green approach

Click Green chemistry has been in place for long as a scientific term without much advantages until recent times when everything wants or needs to be "green". To be scientific, not fancy here, we have try to connect and compare these two, using the "Principles of Green Chemistry", by Anastas and Warner [9].

Prevention: Huisgen cycloadditon addition reaction and high yielding.

Atom economy: Click chemistry synthetic methods should be designed to maximize the incorporation of all materials used in the process into the final product.

Designing safer chemicals: Huisgen reaction products should be designed to affect their desired function while minimizing their toxicity.

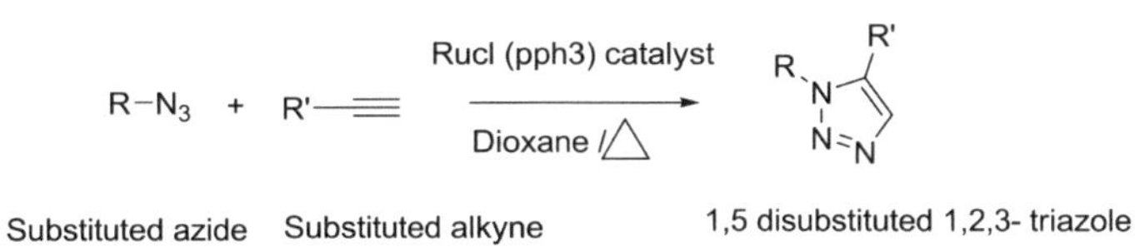

Figure 2. Ruthenium catalyzed azide alkyne cycloaddition.

Safer solvents and auxiliaries: Click chemistry in reaction medium is used water as the solvent.

Design for energy efficiency: Click chemistry is a lot of reactions can be done without much heating.

Reduce derivatives: Click chemistry is a biggest for its superior selectivity and tolerance of most functional groups.

Catalysis: Click chemistry is used for chemical or light catalysts.

Inherently safer chemistry for accident prevention: The use of azides in 1,3 dipolar cycloaddition reaction are minimize the chemical accident.

Advantages of click chemistry:

I. The mixture owns only stable compounds.

II. The reaction owns a high yield.

III. To form a desired product in a simple and quantitative way.

IV. Energetically highly favorable linking reaction.

V. The purification can be done on large scale.

VI. The linkage is chemoselective.

VII. Click reaction must be of wide scope, giving consistently high yields with a variety of starting materials.

VIII. It must be easy to perform, be insensitive to oxygen or water and use only readily available reagents.

IX. Reaction work-up and product isolation must be simple, without requiring chromatographic purification [10].

1.7. Pharmaceutical applications of Triazoles

Heterocyclic compounds containing nitrogen plays an important role in agrochemical and pharmaceuticals. The basic heterocyclic rings present in the various medicinal agents are mainly 1, 2, 3-triazole and 1, 2, 4-triazole [11]. Derivatives of 1, 2, 3-triazole have found to anti-HIV, anti-allergenic, antimicrobial, cytostatic, virostatic, anti-inflammatory and antibacterial [12] activities. Triazoles are also being studied for the treatment of obesity and osteoarthritis.

2. Experimental section

2.1. Synthesis of piperazine using click chemistry

In the present investigation, the synthesis of 1-(3-azidopropyl)-4-(3-chlorophenyl) piperazine were synthesized from 1-(3-chlorophenyl)-4-(3-chloropropyl) piperazine compound which on nucleophilic substitution reaction in the presence of NaN_3 and DMSO at 40–45°C, to afford azide intermediated 1-(3-azidopropyl)-4-(3-chlorophenyl) piperazine in good to excellent yield. We have prepared the piperazine triazole by the Huisgen 1, 3-dipolar cycloaddition reaction of 1-(3-azidopropyl) 4-(3-chlorophenyl) piperazine with various substituted alkynes which was prepared reported method in the presence of Cu (I)-catalyst and we got very high yield. The continued interest for the development of efficient and environmentally friendly procedures for the synthesis of heterocyclic compounds, used copper sulfate with its easy availability, cheap cost and operational simplicity prompted us to explore the synthesis of 1, 3-dipolar cycloaddition reaction.

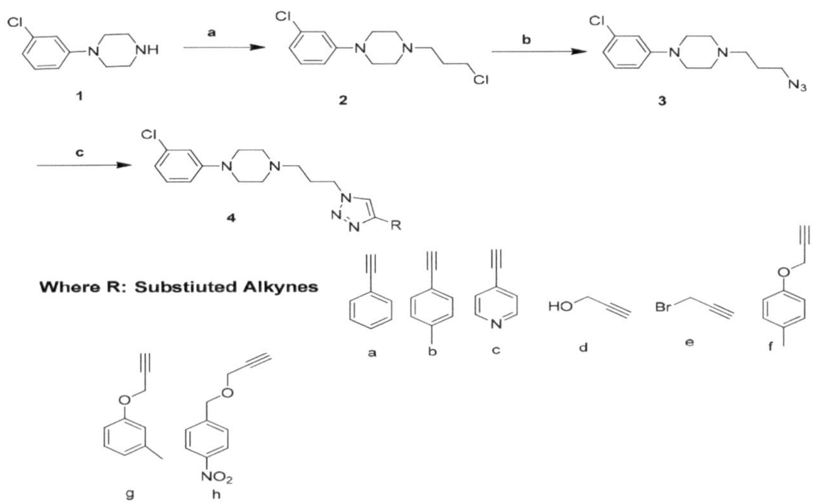

Reaction conditions: **(a)** 1-bromo-3-chloropropane, aq.NaOH, acetone RT, 24 h. **(b)** NaN_3, DMSO, 50–55°C, 4–5 h. **(c)** Click reaction, R-Substituted alkynes, THF: H_2O, Copper sulfate, sodium ascorbate, RT, 10–12 h.

2.2. General procedure

Step (i): To the solution of 1-(3-chlorophenyl)-piperazine **(1)** (0.43 mmol) in water (5 mL) was added sodium hydroxide (1.15 mmol) followed by 1-bromo-3-chloropropane (0.911 mmol) under stirring at 25–30°C. The reaction mixture was further stirred for 24 h at same temperature and progress of reaction was monitored by TLC. After completion of the reaction, the reaction mass was extracted with ethyl acetate. The organic layer was separated and dried over sodium sulfate to obtain pale yellow oily product **(2)** after evaporation of ethyl acetate.

Step (ii): Sodium azide (6.5 mmol) was added to a solution of 1-(3-chlorophenyl)-4-(3-chloropropyl) piperazine (**2**) (5.0 mmol) in 30 mL DMSO under stirring. The reaction mixture was stirred for 3–4 h at 50–55°C. The reaction progress was monitored by TLC. After completion of reaction, the reaction mixture was poured on crushed ice, which was extracted with ethyl acetate. The organic layer was separated and washed with water and brine solution, dried over sodium sulfate to obtain yellow oily crude product. The crude product was purified by column chromatography (ethyl acetate: n-hexane) to obtain pure yellow oily product (**3**).

Step (iii): The azide compound **(3)** (1.0 mmol) and alkyne (1.1 mmol) were dissolved in THF/H_2O (1:1), $CuSO_4 \cdot 5H_2O$ (0.05 mmol) and sodium ascorbate (0.40 mmol). The reaction mixture was stirred for 10 h at room temperature. The progress of reaction was monitored by TLC. After completion of reaction, reaction mixture was poured on crushed ice. The solid obtained was extracted with ethyl acetate. The organic extract was washed with water and brine. The solvent was removed under reduced pressure to afford crude product, which was purified from ethanol to obtain pure compounds.

2.3. Synthesis of quinoxaline by using click chemistry

In the present investigation, the 6-azido-5-bromo quinoxaline were synthesized from 5-bromo quinoxalin-6-amine compound, which on diazotization in the presence of concentrated sulfuric acid, water and sodium nitrite at temperature 0–5°C which undergoes and nucleophilic substitution reaction with sodium azide to afford the 6-azido-5-bromo quinoxaline in good to excellent yield. The quinoxaline 1, 2, 3 triazole derivatives were prepared by the copper catalyzed azide and alkyne cycloaddition reaction of 6-azido-5-bromoquinoxaline with various substituted alkynes were prepared by reported method using copper sulfate and sodium ascorbate in DMF:H_2O as a reaction medium at room temperature to obtain 1,2,3-triazole quinoxaline as shown scheme. The synthesized products were obtained in good to excellent yields. The progress of the reaction was monitored by TLC. Some synthesized compounds were characterized by IR, ^1H NMR, ^{13}C NMR and Mass spectroscopy methods. Some synthesized compounds are antioxidant, antibacterial and antifungal activities have been evaluated.

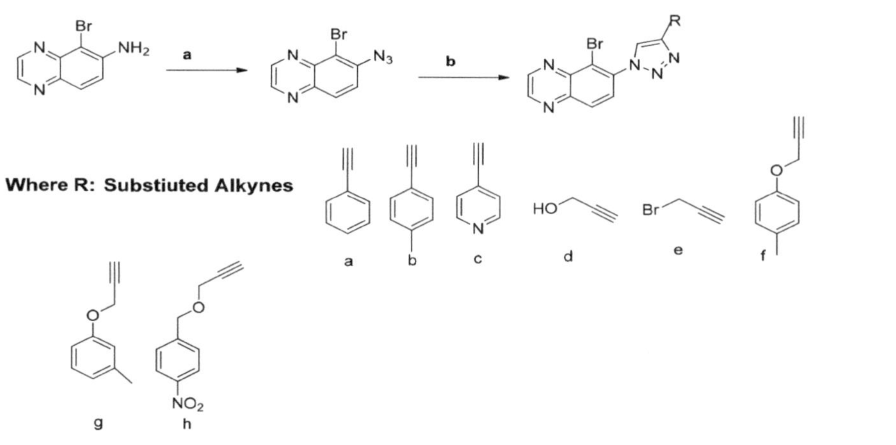

Reaction conditions: (a) H_2O, H_2SO_4, $NaNO_2$, NaN_3, 0–5°C to RT, 3 h. **(b)** Click reaction, RTHF: H_2O, Copper sulfate, sodium ascorbate, RT, 10–13 h.

2.4. General procedure

Step (i): A solution of sodium nitrate (3.13 mmol) in water (8 mL) was added dropwise to a solution of 4-amino-5-bromoquinoxaline **(i)** (2.45 mmol) in water (5 mL) and concentrated

H_2SO_4 (3 mL) at 0°C over 5 min. The reaction mixture was stirred at 0°C for 30 min. Then added solution of sodium azide (4.40 mmol) in water (5 mL). The solution was allowed to attain room temperature and keep stirring for 5 h. The progress of reaction was monitored by TLC. The reaction mass was precipitated, filtered and washed with water. Brown colored crude product was recrystallized from aqueous methanol to obtain pure azide compounds.

Step (ii): The azide compound (1.0 mmol) and alkyne (1.1 mmol) were dissolved in DMF/H_2O (9:1). To this solution, $CuSO_4{\cdot}5H_2O$ (0.05 mmol) and sodium ascorbate (0.40 mmol) were added. The reaction mixture was stirred for 11 h at room temperature. The progress of reaction was monitored by TLC. After completion of reaction, reaction mixture was poured on crushed ice. The solid product was extracted with ethyl acetate. The organic extract was washed with water and brine. The solvent was removed under reduced pressure to afford crude product, which was recrystallized from methanol to obtain pure compound.

2.5. Synthesis of 1, 2, 3-triazoles by one pot method by using click chemistry

The chemoselective azide and alkyne cycloadditions at room temperature in organic medium. K. Barry Sharpless and co-workers have reported a high yielding synthesis of triazoles using a Cu (I)-catalyst with an excellent 1, 4-regioselectivity. The resulting "clicked" products can even be obtained via in situ generation of the corresponding organic azides, halides, NaN_3 in the presence of an alkyne and a copper catalyst, avoiding the need to handle azides.

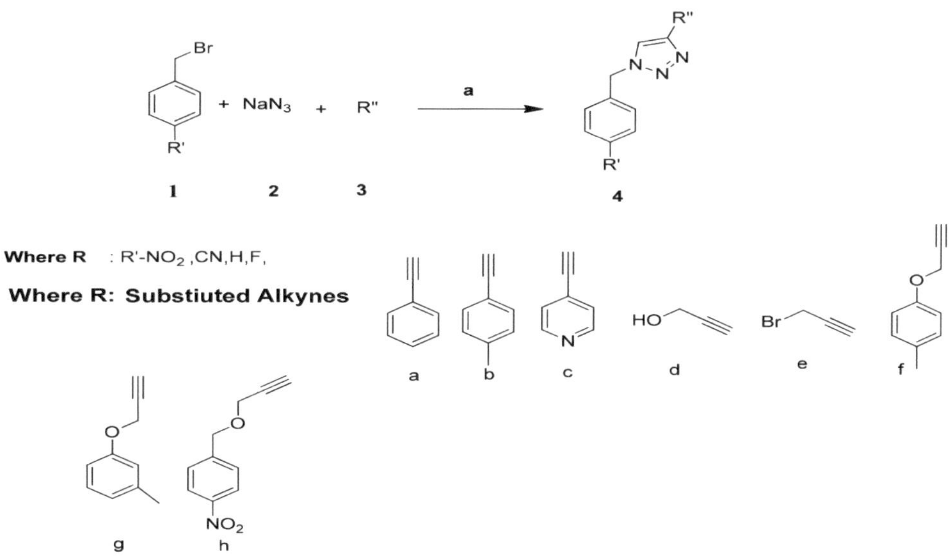

Reagent and conditions: (a) NaN_3, CUI (10%), PEG-400.

2.6. General procedure

Substituted halide (1.0 mmol), sodium azide (1.4 mmol) and Substituted alkynes (1.1 mmol) were charged into the single neck R.B. flask contains polyethylene glycol-400 (5 mL). Catalytic amount of copper iodide (10 mol %) were added into the reaction mixture and maintain it for 6 h at 40–45°C. The progress of reaction was monitored by TLC. After completion of reaction, the mixture was poured on crushed ice. The isolated product was extracted with ethyl acetate. The organic layer was separated and washed with water and brine solution. The solvent was removed under reduced pressure and the isolated crude product was recrystallized from ethanol to obtain pure compounds.

2.7. Synthesis of 1, 2, 3-bistriazoles by using click chemistry

Present investigation in the synthesis of 1, 2, 3-bistriazole, the most widely used is the Cu (I)-catalyzed 1, 3 dipolar cycloaddition reaction in which the condensation of a bis-halide with an substituted alkynes were prepared by reported method in the presence of NaN_3, Na_2CO_3, $CuSO_4 \cdot 5H_2O$, ascorbic acid, DMF: H_2O, 15–20 h, r.t. We have synthesized the 1, 2, 3-bistriazole derivatives by changing the pharmacophore and changing the position of the pharmacophore on substituted alkynes. These synthesized new drug scaffold. The synthesized compounds were evaluated for antibacterial activity and carcinogenicity study.

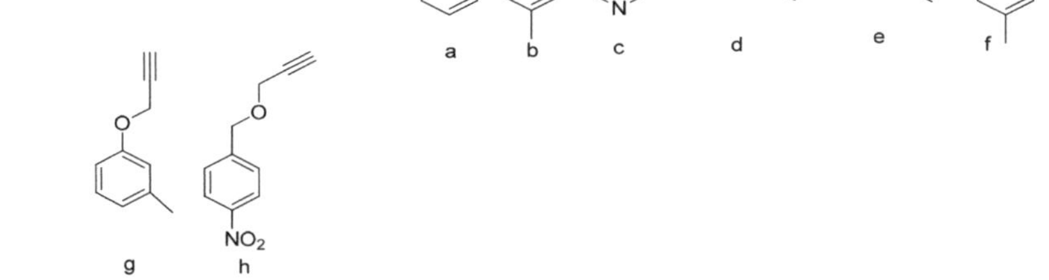

Reagents and condition: (a) NaN_3, DMSO, 45–50°C, 4–5 h. **(b)** CuI, DIPEA, DMF, 5–6 h, 55–60°C.

2.8. General procedure

To a stirred solution of 1, 3-dibromopropane (1.5 mmol) in DMF: H_2O (4:1) 15 mL; NaN_3 (3.2 mmol), Na_2CO_3 (2.2 mmol), $CuSO_4 \cdot 5H_2O$ (0.6 mmol), ascorbic acid (2.2 mmol) and phenyl acetylene (3.1 mmol) were added. The reaction mixture was stirred at room temperature for 20 h. The progress of reaction monitored by TLC. Then, aqueous NH_4OH and CH_2Cl_2 were added in the reaction mixture and the organic layer was separated and washed with water, brine solution and dried by $MgSO_4$. The organic solvent was evaporated under reduced pressure to get crude product. The isolated crude product was recrystallized from ethanol to obtain pure compound.

3. Result and discussion

IR spectra of azide showed characteristic band at near region 2113 cm^{-1} due to ($-N_3$) stretching vibrations. IR spectrums in azido and alkyne peak are disappeared to confirmed 1, 2, 3-triazole formation, of compounds. These assignments are in agreement with those observed by several research groups.

^{1}H NMR spectra of compounds were studied in $CDCl_3$ and DMSO-d_6 showed spectra the proton in triazole ring significantly observed in the region at δ 8.62–7.81 ppm and adjacent sp^2 hybridized carbon of that proton at δ 129.68–127.86 ppm in ^{13}C NMR. These findings are in agreements with those observed by different workers.

The mass spectra of corresponding 1, 2, 3-triazol-1-yl piperazine show their molecular formula weight and found to be in agreement with the literature.

4. Conclusion

We have successfully introduced azide-alkyne 1, 3-dipolar cycloaddition reaction in heterocyclic chemistry. Due to the presence of triazole it observed that enhancing the bioactivity of basic moieties at different heterocycles. We have concluded that a series of novel 1, 2, 3-triazol-1-yl piperazine, quinoxaline, one pot 1,2,3-triazole and bistriazole derivatives by using click chemistry. These derivatives we have achieved by using Husign 1, 3-dipolar cycloaddition which is green chemistry approach because of high yield, high purity, stereo specific, simple to perform, using green solvents.

Acknowledgements

I gratefully acknowledge my deep gratitude to the **Prof. N. N. Maldar**, Vice Chancellor, Solapur University and **Prof. R. B. Bhosale** research guide & Director, School of Chemical Sciences, Solapur University Solapur for providing necessary laboratory facilities.

Author details

Sachin P. Shirame* and Raghunath B. Bhosale

*Address all correspondence to: sachinshirame@gmail.com

School of Chemical Sciences, Solapur University, Solapur, Maharashtra State, India

References

[1] Kolb HC, Finn MG, Sharpless KB. Angewandte Chemie, International Edition. 2001;**40**: 2004-2021

[2] Yigit S, Sanyal R, Sanyal A. Chemistry – An Asian Journal. 2011;**6**:2648-2659

[3] Lee LV, Mitchell ML, Huang SJ, Fokin VV, Sharpless KB, Wong CH. Journal of the American Chemical Society. 2003;**125**:9588

[4] Feldman WP, Nugent AK, Hawker AK, Scheel CJ, Voit A, Pyun B, Fréchet J, Sharpless KB, Fokin VV. Angewandte Chemie, International Edition. 2004;**43**:3928-3932

[5] Rostovtsev VV, Green LG, Fokin VV, Sharpless KB. Angewandte Chemie, International Edition. 2002;**41**:2596

[6] Domling A. Chemical Reviews. 2006;**106**:17

[7] Punna S, Holzer P, Pherson AK, Sharpless KB, Fokin VV, Finn MG. Journal of Polymer Science, Part A. 2004;**42**:4392-4403

[8] Tornøe CW, Meldal M. Proc. Am. Pept. Symp. San Diego; 2001. pp. 263-264

[9] Peabody OK, Peterson MJ, Warner J. Green Chemistry: Terminology and principles (Letter). Environmental Health Perspectives. 2009;**117**:A385-A386

[10] Dolitzky BZ. Int. Appl. WO 00/63, 185. Chemical Abstracts. 2000;**133**:321-901

[11] Roderick WR, Platte HJ, Pollard CB. Journal of Medicinal Chemistry. 1966;**9**:181-185

[12] Coats SJ, Link JS, Gauthier D, Hlasta. Journal of organic Letters. 2005;**7**:1469-1472